Production-Ready Applied Deep Learning

Learn how to construct and deploy complex models in PyTorch and TensorFlow deep learning frameworks

Tomasz Palczewski

Jaejun (Brandon) Lee

Lenin Mookiah

BIRMINGHAM—MUMBAI

Production-Ready Applied Deep Learning

Publishing Product Manager: Ali Abidi
Senior Editor: Nazia Shaikh
Content Development Editor: Shreya Moharir
Technical Editor: Rahul Limbachiya
Copy Editor: Safis Editing
Project Coordinator: Farheen Fathima
Proofreader: Safis Editing
Indexer: Rekha Nair
Production Designer: Aparna Bhagat
Marketing Coordinators: Shifa Ansari and Abeer Dawe

First published: September 2022

Production reference: 1260822

Published by Packt Publishing Ltd.
Livery Place
35 Livery Street
Birmingham
B3 2PB, UK.

ISBN 978-1-80324-366-5

www.packt.com

To Sylwia, Anna, and Matt – my loves, my life.
To my Mom, my brother Piotr, and my family.

- Tomasz

To my parents, Changhee and Kyung Ja, for always loving and supporting me.

- Jaejun

To my mom, Chendurkani, for her unconditional support and encouragement.

- Lenin

Finally, we would like to dedicate this book to self-motivated and
value-driven individuals who put their time into learning
new technologies to make the world more exciting.

Contributors

About the authors

Tomasz Palczewski is currently working as a staff software engineer at **Samsung Research America** (**SRA**). He has a Ph.D. in physics and an eMBA degree from Quantic. His zeal for getting insights out of large datasets using cutting-edge techniques led him to work across the globe at CERN (Switzerland), LBNL (Italy), J-PARC (Japan), University of Alabama (US), and the University of California, Berkeley (US). In 2016, he was deployed to the South Pole to calibrate the world's largest neutrino telescope. At some point, he decided to pivot his career and focus on applying his skills in industry. Currently, Dr. Palczewski works on modeling user behavior and creating value for advertising and marketing verticals by deploying **machine learning** (**ML**), deep learning, and statistical models at scale.

I had the idea of writing a book that my younger self would appreciate. The book would show different aspects of production-ready deep learning. I am grateful that Jaejun and Lenin were excited about the idea and joined the project. Without their help, this would not have turned out as it did. Finally, I would like to thank my wife for all her support.

Jaejun (Brandon) Lee is currently working as an AI research lead at RoboEye.ai, integrating cutting-edge algorithms in computer vision and AI into industrial automation solutions. He obtained his master's degree from the University of Waterloo with research focused on **natural language processing** (**NLP**), specifically speech recognition. He has spent many years developing a fully productionized yet open source wake word detection toolkit with a web browser deployment target, Howl. Luckily, his effort has been picked up by Mozilla's Firefox Voice and it is actively providing a completely hands-free experience to many users all over the world.

I would like to thank Tomasz for offering this remarkable opportunity to become an author. Next, I am really grateful to Lenin for sharing his knowledge of data engineering throughout our journey. Lastly, I would like to thank Erica for her encouragement.

Lenin Mookiah is a machine learning engineer who has worked with reputed tech companies – Samsung Research America, eBay Inc., and Adobe R&D. He has worked in the technology industry for over 11 years in various domains: banking, retail, eDiscovery, and media. He has played various roles in the end-to-end productization of large-scale machine learning systems. He mainly employs the big data ecosystem to build reliable feature pipelines that data scientists consume. Apart from his industrial experience, he researched anomaly detection in his Ph.D. at Tennessee Tech University (US) using a novel graph-based approach. He studied entity resolution on social networks during his master's at Tsinghua University, China.

Working with Tomasz and Jaejun was very exciting. I sincerely thank both for the collaboration on this book. I have learned many aspects of data science from both.

About the reviewers

Utkarsh Srivastava is an AI/ML professional, trainer, YouTuber, and blogger. He loves to tackle and develop ML, NLP, and computer vision algorithms to solve complex problems. He started his data science career as a blogger of his own blog (`datamahadev.com`) and YouTube channel (`datamahadev`), followed by working as a senior data science trainer in an institute in Gujarat. Additionally, he has trained and counseled 1,000+ working professionals and students in AI/ML. Utkarsh has successfully completed 40+ freelance training and development work/projects in data science and analytics, AI/ML, Python development, and SQL. He hails from Lucknow and is currently settled in Bangalore, India, as an analyst at Deloitte USI Consulting.

I would like to thank my mother, Mrs. Rupam Srivastava, for her continuous guidance and support throughout my hardships and struggles. Thanks also to the Supreme Para-Brahman.

Neeraj Jhaveri is a cloud solution architect at Microsoft with expertise in providing data and AI solutions. He has around 20 years of IT experience. Over the last decade, working on data and analytics, he has provided AI architect solutions on Azure. Using Azure ML and Cognitive Services, he has helped customers move to Azure using the latest technologies. He received a master's degree in computer science from NYIT. He provides frequent tech talks for the fast-tracking implementation of AI solutions in Azure.

Pooya Rezaei is an ML software engineer at Google using machine learning to estimate offline conversions from Google Ads. Previously, he was an ML engineer at Iterable for two years optimizing their email marketing automation platform to maximize reach. He received a B.Sc. from the University of Tehran, an M.Sc. from the Sharif University of Technology, and a Ph.D. from the University of Vermont, all in electrical and computer engineering.

Table of Contents

Part 1 – Building a Minimum Viable Product

1

2

3

Developing a Powerful Deep Learning Model   49

4

Experiment Tracking, Model Management, and Dataset Versioning 95

Part 2 – Building a Fully Featured Product

5

Data Preparation in the Cloud   113

6

Efficient Model Training   147

7

Revealing the Secret of Deep Learning Models   185

Part 3 – Deployment and Maintenance

8

Simplifying Deep Learning Model Deployment   201

9

Scaling a Deep Learning Pipeline   209

10

Improving Inference Efficiency   237

Preface

With the growing interest in **artificial intelligence** (**AI**), there are millions of resources introducing various **deep learning** (**DL**) techniques for a wide range of problems. They might be sufficient to get you a data scientist position that many of your friends dream of. However, you will soon find out that the real difficulty with DL projects is not only selecting the right algorithm for the given problem but also efficiently preprocessing the necessary data in the right format and providing a stable service.

This book walks you through every step of a DL project. We start from a proof-of-concept model written in a notebook and transform the model into a service or application with the goal of maximizing user satisfaction upon deployment. Then, we use **Amazon Web Services** (**AWS**) to efficiently provide a stable service. Additionally, we look at how to monitor a system running a DL model after deployment, closing the loop completely.

Throughout the book, we focus on introducing various techniques that engineers at the frontier of the technology use daily to meet strict service specifications.

By the end of this book, you will have a broader understanding of the real difficulties in deploying DL applications at scale and will be able to overcome these challenges in the most efficient and effective way.

Who this book is for

Machine learning engineers, deep learning specialists, and data scientists will find this book helpful in closing the gap between the theory and application with detailed examples. Beginner-level knowledge in machine learning or software engineering will help you grasp the concepts covered in this book easily.

What this book covers

Chapter 1, *Effective Planning of Deep Learning-Driven Projects*, is all about how to prepare a DL project. We introduce various terminologies and techniques used in project planning and describe how to construct a project playbook that summarizes the plan.

Chapter 2, *Data Preparation for Deep Learning Projects*, describes the first steps of a DL project, data collection and data preparation. In this chapter, we cover how to prepare a notebook setting for the project, collect the necessary data, and process it effectively for training a DL model.

Chapter 3, Developing a Powerful Deep Learning Model, explains the theory behind DL and how to develop a model using the most popular frameworks: PyTorch and TensorFlow.

Chapter 4, Experiment Tracking, Model Management, and Dataset Versioning, introduces a set of useful tools for experiment tracking, model management, and dataset versioning, which enables effective management of a DL project.

Chapter 5, Data Preparation in the Cloud, focuses on using AWS for scaling up a data processing pipeline. Specifically, we look at how to set up and schedule **extract, transform, and load** (**ETL**) jobs in a cost-efficient manner.

Chapter 6, Efficient Model Training, starts by describing how to configure TensorFlow and PyTorch training logic to utilize multiple CPU and GPU devices on different machines. Then, we look at tools developed for distributed training: SageMaker, Horovod, Ray, and Kubeflow.

Chapter 7, Revealing the Secret of Deep Learning Models, introduces hyperparameter tuning, the most standard process of finding the right training configuration. We also cover **Explainable AI**, a set of processes and methods for understanding what DL models do behind the scenes.

Chapter 8, Simplifying Deep Learning Model Deployment, describes how you can utilize **open neural network exchange** (**ONNX**), a standard file format for machine learning models, to convert models for various frameworks, which helps in separating the model development from model deployment.

Chapter 9, Scaling a Deep Learning Pipeline, covers the two most popular AWS features designed for deploying a DL model as an inference endpoint: **Elastic Kubernetes Service** (**EKS**) and SageMaker.

Chapter 10, Improving Inference Efficiency, introduces techniques for improving the inference latency upon deployment while maintaining the original performance as much as possible: network quantization, weight sharing, network pruning, knowledge distillation, and network architecture search.

Chapter 11, Deep Learning on Mobile Devices, describes how to deploy TensorFlow and PyTorch models on mobile devices using TensorFlow Lite and PyTorch Mobile, respectively.

Chapter 12, Monitoring Deep Learning Endpoints in Production, explains existing solutions for monitoring a system running a DL model in production. Specifically, we discuss how to integrate CloudWatch into endpoints running on SageMaker and EKS clusters.

Chapter 13, Reviewing the Completed Deep Learning Project, covers the last phase of a DL project, the reviewing process. We describe how to effectively evaluate a project and prepare for the next project.

To get the most out of this book

Even though we will interact with many tools throughout our journey, all the installation instructions are included in the book and the GitHub repository. The only thing you will need to prepare prior to reading this book would be an AWS account. AWS provides a Free Tier (`https://aws.amazon.com/free`), which should be sufficient to get you started.

Software/hardware covered in the book	Operating system requirements
TensorFlow	Windows, macOS, or Linux
PyTorch	
Docker	
Weights & Biases, MLflow, and DVC	
ELI5 and SHAP	
Ray and Horovod	
AWS SageMaker	
AWS EKS	

If you want to try running the samples in the book, we advise you to use the complete versions from either our repository or the official documentation pages as the versions in the book may have some components missing to enhance the delivery of the contents.

Download the example code files

You can download the example code files for this book from GitHub at `https://github.com/PacktPublishing/Production-Ready-Applied-Deep-Learning`. If there's an update to the code, it will be updated in the GitHub repository.

We also have other code bundles from our rich catalog of books and videos available at `https://github.com/PacktPublishing/`. Check them out!

Download the color images

We also provide a PDF file that has color images of the screenshots and diagrams used in this book. You can download it here: `https://packt.link/fUhAv`.

Conventions used

There are a number of text conventions used throughout this book.

`Code in text`: Indicates code words in text, database table names, folder names, filenames, file extensions, pathnames, dummy URLs, user input, and Twitter handles. Here is an example: "Mount the downloaded `WebStorm-10*.dmg` disk image file as another disk in your system."

A block of code is set as follows:

```
html, body, #map {
  height: 100%;
  margin: 0;
  padding: 0
}
```

When we wish to draw your attention to a particular part of a code block, the relevant lines or items are set in bold:

```
[default]
exten => s,1,Dial(Zap/1|30)
exten => s,2,Voicemail(u100)
exten => s,102,Voicemail(b100)
exten => i,1,Voicemail(s0)
```

Any command-line input or output is written as follows:

```
$ mkdir css
$ cd css
```

Bold: Indicates a new term, an important word, or words that you see onscreen. For instance, words in menus or dialog boxes appear in **bold**. Here is an example: "Select **System info** from the **Administration** panel."

> **Tips or important notes**
> Appear like this.

Get in touch

Feedback from our readers is always welcome.

General feedback: If you have questions about any aspect of this book, email us at customercare@packtpub.com and mention the book title in the subject of your message.

Errata: Although we have taken every care to ensure the accuracy of our content, mistakes do happen. If you have found a mistake in this book, we would be grateful if you would report this to us. Please visit www.packtpub.com/support/errata and fill in the form.

Piracy: If you come across any illegal copies of our works in any form on the internet, we would be grateful if you would provide us with the location address or website name. Please contact us at copyright@packt.com with a link to the material.

If you are interested in becoming an author: If there is a topic that you have expertise in and you are interested in either writing or contributing to a book, please visit authors.packtpub.com.

Share Your Thoughts

Once you've read Production-Ready Applied Deep Learning, we'd love to hear your thoughts! Scan the QR code below to go straight to the Amazon review page for this book and share your feedback.

https://packt.link/r/1-803-24366-X

Your review is important to us and the tech community and will help us make sure we're delivering excellent quality content.

Part 1 –
Building a Minimum
Viable Product

AI projects begin with planning and understanding the difficulty of the given problem. Once the scope of the project is clearly defined, the next step is to create a **Minimum Viable Product (MVP)**. For a project based on **deep learning**, this process involves preparing a set of data and exploring various model architectures to come up with a working solution to the problem. In this first part of the book, we explain how you can carry out the aforementioned steps efficiently by exploiting the various resources available.

This part comprises the following chapters:

- *Chapter 1, Effective Planning of Deep Learning-Driven Projects*
- *Chapter 2, Data Preparation for Deep Learning Projects*
- *Chapter 3, Developing a Powerful Deep Learning Model*
- *Chapter 4, Experiment Tracking, Model Management, and Dataset Versioning*

1

Effective Planning of Deep Learning-Driven Projects

In the first chapter of the book, we would like to introduce what **deep learning** (**DL**) is and how DL projects are typically carried out. The chapter begins with an introduction to DL, providing some insight into how it shapes our daily lives. Then, we will shift our focus to DL projects by describing how they are structured. Throughout the chapter, we will put extra emphasis on the first phase, project planning; you will learn key concepts such as the comprehension of business objectives, how to define appropriate evaluation metrics, identification of stakeholders, resource planning, and the difference between a **minimum viable product** (**MVP**) and a **fully featured product** (**FFP**). By the end of this chapter, you should be able to construct a DL project playbook that clearly describes the goal of the project, milestones, tasks, resource allocation, and its timeline.

In this chapter, we're going to cover the following main topics:

- What is DL?
- Understanding the role of DL in our daily lives
- Overview of DL projects
- Planning a DL project

Technical requirements

You can download the supplemental material for this chapter from the following GitHub link:

https://github.com/PacktPublishing/Production-Ready-Applied-Deep-Learning/tree/main/Chapter_1

What is DL?

It has only been a decade since DL emerged but it has rapidly started playing an important role in our daily lives. Within the field of **artificial intelligence** (**AI**), a popular set of methods categorized as **machine learning** (**ML**) exists. By extracting meaningful patterns from historical data, the goal of ML is to build a model that makes sensible predictions and decisions for newly collected data. DL is an ML technique that exploits **artificial neural networks** (**ANNs**) to capture a given pattern. *Figure 1.1* presents the key components of the AI evolution that started around 1950s, with Alan Turing conducting discussions about intelligent machines, among other godfathers of the field. While various ML algorithms have been introduced sporadically since the advent of AI, it actually took another decades for the field to bloom. Similarly, it has only been about a decade since DL has became the main stream because many of the algorithms in this field require extensive computational power.

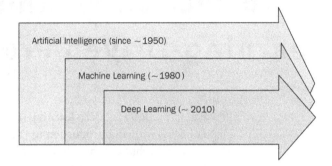

Figure 1.1 – A history of AI

As shown in *Figure 1.2*, the key advantage of DL comes from ANNs, which enable the automatic selection of necessary features. Similar to the way that human brains are structured, ANNs are also made up of components called **neurons**. A group of neurons forms a **layer** and multiple layers are linked together to form a **network**. This kind of architecture can be understood as multiple steps of nested instructions. As the input data passes through the network, each neuron extracts different information, and the model is trained to select the most relevant features for the given task. Considering the role of neurons as building blocks for pattern recognition, deeper networks generally lead to greater performance, as they learn the details better:

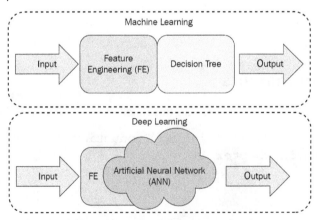

Figure 1.2 – The difference between ML and DL

While typical ML techniques require features to be manually selected, DL learns to select important features on its own. Therefore, it can potentially be adapted to a broader range of problems. However, this advantage does not come for free. In order to train a DL model properly, the datasets for training need to be large and diverse enough, which leads to an increase in training time. Interestingly, **graphics processing unit (GPU)** has played a major role in reducing the training time. While a **central processing unit (CPU)** demonstrates its effectiveness in carrying out complex computations with its broader instruction sets, a GPU is more suitable for processing simpler but larger computations with its massive parallelism. By exploiting such an advantage in the matrix multiplications that the DL model heavily depends on, GPU has become a critical component within DL.

As we are still in the early stages of the AI era, chip technology is evolving continuously, and it is not yet clear how these technologies will change in the future. It is worth mentioning that new designs come from start-ups as well as big tech companies. This ongoing race clearly shows that more and more products and services based on AI will be introduced. Considering the growth in the market and job opportunities, we believe that it is a great time to learn about DL.

> **Things to remember**
>
> a. DL is an ML technique that exploits ANNs for pattern recognition.
>
> b. DL is highly flexible because it selects the most relevant features automatically for the given task throughout training.
>
> c. GPUs can speed up DL model training with its massive parallelism.

Now that we understand what DL is at a high level, we will describe how it shapes our daily lives in the next section.

Understanding the role of DL in our daily lives

By exploiting the flexibility of DL, researchers have made remarkable progress in the domains in which traditional ML techniques have shown limited performance (see *Figure 1.3*). The first flag has been planted in the field of **computer vision (CV)** for digit recognition and object detection tasks. Then, DL has been adopted for **natural language processing (NLP)**, showing meaningful progress in translation and speech recognition tasks. It also demonstrates its effectiveness in **reinforcement learning (RL)** as well as **generative modeling**.

The list of papers linked in the *Further reading* section in this chapter summarizes popular use cases of DL.

Following diagram shows various applications of DL:

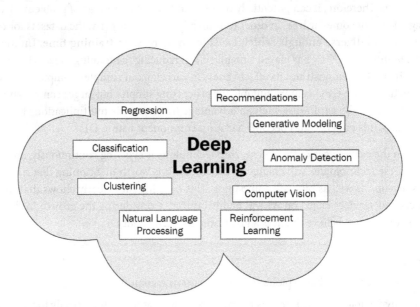

Figure 1.3 – Applications of DL

However, integrating DL into an existing technology infrastructure is not an easy task; difficulties can arise from various aspects, including but not limited to budget, time, as well as talent. Therefore, a thorough understanding of DL has become an essential skill for those who manage DL projects: project managers, technology leads, as well as C-suite executives. Furthermore, the knowledge in this fast-growing field will allow them to find future opportunities in their specific verticals and across the organization, as people working on AI projects actively gather new knowledge to derive innovative and competitive advantages. Overall, an in-depth understanding of DL pipelines and developing production-ready outputs allows managers to execute projects better by effectively avoiding commonly known pitfalls.

Unfortunately, DL projects are not yet in a plug-and-play state. In many cases, they involve extensive research and adjustment phases, which can quickly drain the available resources. Above all, we have noticed that many companies struggle to move from **proof of concept** to production because critical decisions are made by the few who only have a limited understanding of the project requirements and DL pipelines. With that being said, our book aims to provide a complete picture of a DL project; we will start with project planning, and then discuss how to develop MVPs and FFPs, how to utilize cloud services to scale up, and finally, how to deliver the product to targeted users.

> **Things to remember**
>
> a. DL has been applied to many problems in various fields, including but not limited to CV, NLP, RL, and generative modeling.
>
> b. An in-depth understanding of DL is crucial for those leading DL projects, regardless of their position or background.
>
> c. This book provides a complete picture of a DL project by describing how DL projects are carried out from project planning to deployment.

Next, we will take a closer look at how DL projects are structured.

Overview of DL projects

While DL and other software engineering projects have a lot in common, DL projects emphasize planning, due to the extensive need for resources, mainly coming from the complexity of the models and the high volume of data involved. In general, DL projects can be split into the following phases:

1. Project planning
2. Building MVPs
3. Building FFPs
4. Deployment and maintenance
5. Project evaluation

In this section, we provide high-level overviews of these phases. The following sections cover each phase in detail.

Project planning

As the first step, the project lead must clearly define what needs to be achieved by the project and understand groups that can affect or be affected by the project. The evaluation metrics need to be defined and agreed upon, as they will be revisited during project evaluation. Then, the team members group together to discuss individual responsibilities and achieve business objectives using available resources. This process naturally leads to a timeline, an estimate of how long the project would take. Overall, project planning should result in the generation of a document called a playbook, which includes a thorough description of how the project will be carried out and evaluated.

Building minimum viable products

Once the direction is clear for everyone, the next step is to build an MVP, a simplistic version of the target deliverable that showcases the project's value. Another important aspect of this phase is to understand the project's difficulties and reject paths with greater risks or less promising outcomes by following the *fail fast, fail often* philosophy. Therefore, data scientists and engineers typically work with partially sampled datasets in development settings and ignore insignificant optimizations.

Building fully featured products

Once the feasibility of the project has been confirmed by the MVP, it must be packaged into an FFP. This phase aims to polish up the MVP to build a production-ready deliverable with various optimizations. In the case of DL projects, additional data preparation techniques are introduced to improve the quality of input data, or the model pipeline gets augmented slightly for greater model performance. Additionally, the data preparation pipeline and model training pipeline may be migrated to the cloud, exploiting various web services for higher throughput and quality. In this case, the whole pipeline often gets reimplemented using different tools and services. This book focuses on **Amazon Web Services (AWS)**, the most popular web service for handling high volumes of data and expensive computations.

Deployment and maintenance

In many cases, the deployment settings are different from the development settings. Therefore, different sets of tools are often involved when moving an FFP to production. Furthermore, deployment may introduce problems that weren't visible during development, which mainly arise as a result of limited computational resources. Consequently, many engineers and scientists spend additional time improving the user experience during this phase. Most people believe that deployment is the last step. However, there is one more step: maintenance. The quality of data and model performance needs to be monitored consistently to provide stable services to targeted users.

Project evaluation

In the last phase, project evaluation, the team should revisit the discussions made during project planning to evaluate whether the project has been carried out successfully or not. Furthermore, the details of the project need to be recorded, and potential improvements must be discussed so that the next projects can be achieved more efficiently.

Things to remember

a. The phases within DL projects are split into project planning, building MVPs, building FFPs, deployment and maintenance, and project evaluation.

b. During the project planning phase, the project goal and evaluation metrics are defined, and the team discusses an individual's responsibility, available resources, and the timeline for the project.

c. The purpose of building an MVP is to understand the difficulties of the project and reject paths that pose greater risks or offer less promising outcomes.

d. The FFP is a production-ready deliverable that is an optimized version of the MVP. The data preparation pipeline and model training pipeline may be migrated to the cloud, exploiting various web services for higher throughput and quality.

e. Deployment settings often provide limited computational resources. In this case, the system needs to be tuned to provide stable services to target users.

f. Upon the completion of the project, the team needs to revisit the timeline, assigned responsibilities, and business requirements to evaluate the success of the project.

In the following section, we will walk you through how to plan a DL project properly.

Planning a DL project

Every project starts with planning. Throughout the planning, the purpose of the project needs to be clearly defined, and key members should have a thorough understanding of the available resources that can be allocated to the project. Once team members and stakeholders are identified, the next step is to discuss each individual's responsibility and create a timeline for the project.

This phase should result in a well-documented project playbook that precisely defines business objectives and how the project will be evaluated. A typical playbook contains an overview of key deliverables, a list of stakeholders, a **Gantt chart** defining steps and bottlenecks, definitions of responsibilities, timelines, and evaluation criteria. In the case of highly complex projects, following the **Project Management Body of Knowledge (PMBOK®)** Guide (`https://www.pmi.org/pmbok-guide-standards/foundational/pmbok`) and considering every knowledge domain (for example, integration management, project scope management, and time management) are strongly recommended. Of course, other project management frameworks exist, such as **PRINCE2** (`https://www.prince2.com/usa/what-is-prince2`), which can provide a good starting point. Once the playbook is constructed, every stakeholder must review and revise it until everyone agrees with the contents.

In real life, many people underestimate the importance of planning. Especially in start-ups, engineers are eager to dive into MVP development and spend minimal time planning. However, it is especially dangerous to do so in the case of DL projects because the training process can quickly drain all the available resources.

Defining goal and evaluation metrics

The very first step of planning is to understand what purpose the project serves. The goal might be developing a new product, improving the performance of an existing service, or saving on operational costs. The motivation of the project naturally helps define the evaluation metrics.

In the case of DL projects, there are two types of evaluation metrics: business-related metrics and model-based metrics. Some examples of business-related metrics are as follows: **conversion rate**, **click-through rate (CTR)**, **lifetime value**, **user engagement measure**, **savings in operational cost**, **return on investment (ROI)**, and **revenue**. These are commonly used in advertising, marketing, and product recommendation verticals.

On the other hand, model-based metrics include **accuracy, precision, recall, F1-score, rank accuracy metrics, mean absolute error (MAE), mean squared error (MSE), root-mean-square error (RMSE)**, and **normalized mean absolute error (NMAE)**. In general, tradeoffs can be made between the various metrics. For example, a slight decrease in accuracy may be acceptable if meeting latency requirements is more critical to the project.

Along with the target evaluation metric, which differs from project to project, there are other metrics that are commonly found in most projects. These are due dates and resource usage. The target state must be reached by a certain date using available resources.

The goal and corresponding evaluation metrics need to be fair. If the goal is too hard to achieve, project members can possibly lose motivation. If the metric for the evaluation is not correct, understanding the impact of the project becomes difficult. As a result, it is recommended that the selected evaluation metrics are shared with others and considered fair for everyone.

Project Name: Document Date:	Deep learning–based recommendation system 12/1/21	
General Description: Technologies: Implementation level: Key Goal: Evaluation metrics:	In this project,team X will develop recommendation system for Y vertical to improve current rule-based approach. Python, Spark, TensorFlow, SageMaker Simple / Medium / Hard Increase CTR by 5% comparing to rule-based approach. 1) CTR 2) needs to be able to serve at least X users per Y time 3) cost constraint during development ($X) 4) cost constraint during production ($X)	
Key Features: Additional Features:	Features that team needs to deliver to successfully finishing the project Features that might expand project scope and deliver additional value but are not crucial to succesfuly finish the project	
Stakeholders: Sponsor Project Manager (PM) Technical Product Manager (TPM) Project Team or specific project team groups Internal teams / Stakeholders	Responsibilities: list key responsibilities here, and define form of communication / reporting	Approved by stakeholder: yes/no (adjust until all stakeholders are aligned) ... ...

Figure 1.4 – A sample playbook with the project description section filled out

As shown in *Figure 1.4*, the first section of a playbook begins with a general description, an estimated complexity of the technical aspects, and a list of required tools and frameworks. Next, it clearly describes the objective of the project and how the project will be evaluated.

Stakeholder identification

In the same way that the term stakeholder is used for a business, a stakeholder for a project refers to a person or group who can affect or be affected by the project. Stakeholders can be grouped into two types, internal and external. Internal stakeholders are those that are directly involved in project executions, while external stakeholders may be outside of the circle, supporting the project execution in an indirect way.

Each stakeholder has a different role within the project. First, we'll look at internal stakeholders. Internal stakeholders are the main drivers of the project. Therefore, they work closely together to process and analyze data, develop a model, and build deliverables. *Table 1.1* lists internal stakeholders that are commonly found in DL projects:

Stakeholder	Role
Sponsor	- Initiating the project - Defining a business justification for the project - Canceling the project when it is no longer needed
Project lead	- Motivating team members for the success of the project - Interacting with external stakeholders to make sure that the project is not delayed unexpectedly
Project manager	- Planning, monitoring, and ensuring the stable execution of the project - Analyzing risks - Making sure the project is on schedule
Data engineers	- Preprocessing the necessary data into a form that data scientists can use
Data scientists	- Analyzing the data and developing a model for the project
DevOps	- Migrating the model and data preprocessing logics to the cloud - Supporting software engineers with the deployment of the deliverable
Software engineers	- Developing the necessary tools for the project - Building the deliverable - Deploying the deliverable to the target users

Table 1.1 – Common internal stakeholders for DL projects

On the other hand, external stakeholders often play supportive roles, such as collecting necessary data for the project or providing feedback about the deliverable. In *Table 1.2*, we describe some common external stakeholders for DL projects:

Stakeholder	Role
Data collector	Collecting the raw data that the project depends on
Labeling company	Labeling the raw data for model training
User	Interacting with the deliverable and providing feedback
C-suite executives	Allocating resources to the project

Table 1.2 – Common external stakeholders for DL projects

Stakeholders are described in the second section of a playbook. As shown in *Figure 1.4*, the playbook must list stakeholders and their responsibilities in the project.

Task organization

A milestone refers to a point in a project where a significant event occurs. Therefore, there is a set of requirements leading up to a milestone. Once the requirements are met, a milestone can be claimed to have been reached. One of the most important steps in project planning is defining milestones and their associated tasks. The ordering of tasks that lead to the goal is called the **critical path**. It is worth mentioning that tasks don't need to be tackled sequentially all the time. The understanding of a critical path is important because it allows the team to prioritize tasks appropriately to ensure the success of the project.

In this step, it is also critical to identify **level-of-effort** (**LOE**) activities and supportive activities, which are required for project execution. In the case of software development projects, LOE activities include supplementary tasks, such as setting up Git repositories or reviewing others' merge requests. The following figure (*Figure 1.5*) describes a typical critical path for a DL project. It will be structured differently if the underlying project consists of different tasks, requirements, technologies, and desired levels of detail:

ID	Task Name
1	Project Start
2	Initial Analysis
3	Feature Engineering
4	Creation of train, cross-validation, test sets
5	Model training / proof of concept
6	Model evaluation, Model understanding
7	Hyperparameter tuning
8	Creation of final MVP or service
9	Adjustments to existing environments, automation, data
10	model optimization (i.e. pruning, quantization)
11	Inference tests, test in staging environment
12	Creation of production-ready product or service
13	Setting up maintenance services, model understanding and controlling in production environment
14	Project Closure

Figure 1.5 – A typical critical path for a DL project

Resource allocation

For a DL project, there are two main resources that require explicit resource allocations: human and computational resources. Human resources refer to employees that will actively work on individual tasks. In general, they hold positions in data engineering, data science, DevOps, or software engineering. When people talk about human resources, they often consider headcount only. However, the knowledge and skills that individuals hold are other critical factors. Human resources are closely related to how fast the project can be carried out.

Computational resources refer to hardware and software resources that are allocated to the project. Unlike typical software engineering projects, such as mobile app development or web page development, DL projects require heavy computation and large amounts of data. Common techniques for speeding up the development process involve parallelism or using computationally stronger machines. In some cases, tradeoffs need to be made between them, as a single machine of high computational power can cost more than multiple machines of low computational power.

Overall, novel DL pipelines take advantage of flexible and stateless resources, such as AWS Spot instances with fault-tolerant code. Besides hardware resources, there are frameworks and services that may provide necessary features out of the box. If the necessary service requires a payment, it is important to understand how it can change the project execution and what the demand on human resources would be if the team decided to handle it in-house.

In this step, available resources need to be allocated to each task. *Figure 1.6* lists the tasks described in the previous section and describes the allocated resources, along with estimates of operational costs. Each task has its own risk level indicator. It is designed for a small team of three people working on a simple DL project with limited computational resources on a couple of AWS **Elastic Compute Cloud (EC2)** instances for around 4 to 6 months. Please note that the cost estimation of human resources is not included in the example, as it differs a lot depending on geographic location and team seniority:

ID	Task Name	Optimistic Estimate (O) [days]	Most Likely Estimate (M) [days]	Pessimistic Estimate (P) [days]	Support Type Activities / LOE Estimate [days]	Task Predecessors	Head Count	Team	Start Date	End Date	Risk	Resource Cost	Resource cost estimation method
1	Project Start	1	1	1	0		0	3 BookDL	12/12/21	12/13/21	Low	$0.00 A	
2	Initial Analysis	5	7	9	3		1	3 BookDL	12/13/21	12/23/21	Low	$900.00 A	
3	Feature Engineering	6	8	10	3		2	3 BookDL	12/23/21	1/3/22	Low	$990.00 A	
4	Creation of train, cross-validation, test sets	3	4	5	1		3	1 BookDL	12/24/21	12/29/21	Low	$150.00 A	
5	Model training / proof of concept	7	9	11	4		4	3 BookDL	12/25/21	1/7/22	Low	$1,170.00 A	
6	Model evaluation, Model understanding	3	4	5	1		5	1 BookDL	12/26/21	12/31/21	Medium	$150.00 A	
7	Hyperparameter tuning	6	7	10	2		5	2 BookDL	12/27/21	1/5/22	Medium	$2,700.00 B	
8	Creation of final MVP or service	4	6	8	2		7	3 BookDL	12/28/21	1/5/22	Medium	$720.00 A	
9	Adjustments to existing environments, automation, data	6	8	9	3		8	3 BookDL	12/29/21	1/9/22	High	$990.00 A	
10	model optimization (i.e. pruning, quantization)	3	4	5	2		8	1 BookDL	12/30/21	1/5/22	Medium	$180.00 A	
11	Inference tests, test in staging environment	5	6	7	2		10	3 BookDL	12/31/21	1/8/22	Low	$720.00 A	
12	Creation of production-ready product or service	1	2	3	1		11	1 BookDL	1/1/22	1/4/22	Medium	$90.00 A	
13	Setting up maintenance services, model understanding and controlling in production environment	1	2	3	1		11	1 BookDL	1/2/22	1/5/22	Low	$90.00 A	
14	project clouser	1	1	1	1		13	3 BookDL	1/3/22	1/5/22	Low	$180.00 A	
										Total		$8,030.00	

A   (M estimate + LOE) * Head Count * *8h/day * $5 (*cost of one mp2A.xlarge or p2.8xlarge)
B   (M estimate + LOE) * Head Count * *8h/day * $25 (*cost of one p3.16xlarge)

Figure 1.6 – A sample resource allocation section of a DL project

Before we move on to the next step, we would like to mention that it is important to set aside a portion of the resources as a backup, in case the milestone requires more resources than that have been allocated.

Defining a timeline

Now that we know the available resources, we should be able to construct a timeline for the project. In this step, the team needs to discuss how long each step would take to complete. It is worth mentioning that things don't work out as planned all the time. There will be many difficulties throughout the project that can delay the delivery of the final product.

Therefore, including buffers within the timeline is a common practice in many organizations. It is important that every stakeholder agrees with the timeline. If anyone believes that it's not reasonable, the adjustment needs to be made right away. *Figure 1.7* is a sample Gantt chart with the most likely estimated timeline for the information presented in *Figure 1.6*:

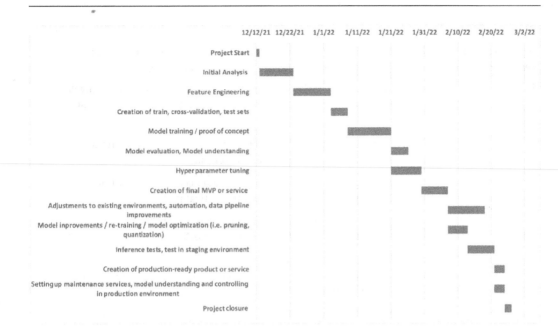

Figure 1.7 – A sample Gantt chart describing the timeline

It is worth mentioning that the chart can also be used to monitor the progress of each task and the overall project. In such cases, additional comments or visualizations summarizing the progress can be attached to each indicator bar.

Managing a project

Another important aspect of a DL project that needs to be discussed during the project planning phase is the process that the team will follow to update other team members and ensure on-time delivery of the project. Out of various project management methodologies, Agile fits perfectly, as it helps to split work into smaller parts and quickly iterate over development cycles until the FFP emerges. As DL projects are generally considered highly complex, it is more convenient to work within short cycles of research, development, and optimization phases. At the end of each cycle, stakeholders review results and adjust their long-term goals. **Agile methodology** is particularly suitable for small teams of experienced individuals. In a typical setting, 2-week sprints are found to be the most effective, especially when the short-term goals are clearly defined.

During a sprint meeting, the team reviews goals from the last sprint and defines goals for the upcoming sprint. It is also recommended to have short daily meetings to go over work performed on the previous day and plan for the upcoming day, as this process can help the team to quickly recognize bottlenecks and adjust their priorities as necessary. Commonly used tools for this process are **Jira**, **Asana**, and **Quickbase**. The majority of the aforementioned tools also support budget management, timeline reviewing, idea management, and resource tracking.

Things to remember

a. Project planning should result in a playbook that clearly describes what purpose the project serves and how the team will move together to reach a particular goal state.

b. The first step of project planning is to define a goal and its corresponding evaluation metrics. In the case of DL projects, there are two types of evaluation metrics: business-related metrics and model-based metrics.

c. A stakeholder refers to a person or a group who can affect or be affected by the project. Stakeholders can be grouped into two types: internal and external.

d. The next stage of project planning is task organization. The team needs to identify milestones, identify tasks, along with LOE activities, and understand the critical path.

e. For DL projects, there are two main resources that require explicit resource allocation: human and computational resources. During resource allocation, it is important to put aside a portion of the resources as a backup.

f. When estimating the timeline for the project, it must be shared within the team, and every stakeholder must agree with the schedule.

g. Agile methodology is a perfect fit for managing DL projects, as it helps to split work into smaller parts and quickly iterate over development cycles.

Summary

This chapter is an introduction to our journey. In the first two sections, we have described where DL sits within the wider picture of AI and how it continually shapes our daily lives. The key takeaways are the fact that DL is highly flexible due to its unique model architecture and the fact that DL has been actively adopted to the domain which traditional ML techniques have failed to demonstrate notable accomplishments.

Then, we have provided a high-level view of the DL project. In general, DL projects can be split into the following phases: project planning, building MVPs, building FFPs, development and maintenance, and project evaluation.

The main contents of this chapter covered the most important step of the DL project: project planning. In this phase, the purpose of the project needs to be clearly defined, along with the evaluation metrics, everyone must have a solid understanding of the stakeholders and their respective roles, and lastly, the tasks, milestones, and timeline need to be agreed upon by the participants. The outcome of this phase would be a well-formatted document called a playbook. In the next chapter, we will learn how to prepare data for DL projects.

Further reading

Here is a list of references that can help you gain more knowledge about the topics that are relevant to this chapter. The following research papers summarize popular use cases of DL:

- CV

    - *Gradient-based learning applied to document recognition* by *LeCun et al.*

    - *ImageNet: A Large-Scale Hierarchical Image Database* by *Deng et al.*

- NLP

    - *A Neural Probabilistic Language Model* by *Bengio et al.*

    - *Speech Recognition with Deep Recurrent Neural Networks* by *Grave et al.*

- RL

    - *An Introduction to Deep Reinforcement Learning* by *François-Lavet et al.*

- Generative modeling

    - *Generative Adversarial Networks* by *Goodfellow et al.*

2

Data Preparation for Deep Learning Projects

The first step in every **machine learning (ML)** project consists of data collection and data preparation. As a subset of ML, **deep learning (DL)** involves the same data processing processes. We will start this chapter by setting up a standard DL Python notebook environment using Anaconda. Then, we will provide concrete examples for collecting data in various formats (JSON, CSV, HTML, and XML). In many cases, the collected data gets cleaned up and preprocessed as it consists of unnecessary information or invalid formats.

The chapter will introduce popular techniques in this domain: filling in missing values, dropping unnecessary entries, and normalizing the values. Next, you will learn common feature extraction techniques: the bag-of-words model, term frequency-inverse document frequency, one-hot encoding, and dimensionality reduction. Additionally, we will present `matplotlib` and `seaborn`, which are the most popular data visualization libraries. Finally, we will cover Docker images, which are snapshots of a working environment that minimizes potential compatibility issues by bundling an application and its dependencies together.

In this chapter, we're going to cover the following main topics:

- Setting up notebook environments
- Data collection, data cleaning, and data preprocessing
- Extracting features from data
- Performing data visualization
- Introduction to Docker

Technical requirements

The supplemental material for this chapter can be downloaded from GitHub at `https://github.com/PacktPublishing/Production-Ready-Applied-Deep-Learning/tree/main/Chapter_2`.

Setting up notebook environments

Python is one of the most popular programming languages that's widely used for data analysis. Its advantage comes from dynamic typing and being compile-free. With its flexibility, it has become the language that data scientists use the most. In this section, we will introduce how to set up a Python environment for a DL project using **Anaconda** and **Preferred Installer Program** (**PIP**). These tools allow you to create a distinct environment for every project while simplifying package management. Anaconda provides a desktop application with a GUI called Anaconda Navigator. We will walk you through how to set up a Python environment and install popular Python libraries for DL projects such as **TensorFlow**, **PyTorch**, **NumPy**, **pandas**, **scikit-learn**, **Matplotlib**, and **Seaborn**.

Setting up a Python environment

Python can be installed from `www.python.org/downloads`. However, Python versions are often available through package managers that are provided by the operating system, such as **Advanced Package Tool** (**APT**) on Linux and **Homebrew** on macOS. Setting up a Python environment begins with installing the necessary packages using PIP, a package management system that allows you to install and manage various Python packages.

Installing Anaconda

When multiple Python projects have been set up on a machine, separating the environments would be ideal as each project may depend on different versions of those packages. Anaconda can help you with environment management as it is designed for both Python package management and environment management. It allows you to create virtual environments where the installed packages are bound to each environment that is currently active. In addition, Anaconda goes beyond the boundaries of Python, allowing users to install non-Python library dependencies.

First things first, Anaconda can be installed from its official website: `www.anaconda.com`. For completeness, we have described the installation process with pictures, at `https://github.com/PacktPublishing/Production-Ready-Applied-Deep-Learning/blob/main/Chapter_2/anaconda/anaconda_graphical_installer.md`.

It can also be installed directly from the Terminal. Anaconda provides installation scripts for each operating system (`repo.anaconda.com/archive`). You can simply download the right version of the script for your system and run it to get Anaconda installed on your machine. As an example, we will describe how to install Anaconda from one of these scripts for macOS: `https://github.com/PacktPublishing/Production-Ready-Applied-Deep-Learning/blob/main/Chapter_2/anaconda/anaconda_zsh.md`.

Setting up a DL project using Anaconda

At this point, you should have an Anaconda environment ready to use. Now, we will create our first virtual environment and install the necessary packages for a DL project:

```
conda create --name bookenv python=3.8
```

You can list the available `conda` environments using the following command:

```
conda info --envs
```

You should see the `bookenv` environment that we created previously. To activate this environment, you can use the following command:

```
conda activate bookenv
```

Similarly, deactivation can be achieved by using the following command:

```
conda deactivate
```

Installing a Python package can be done through either `pip install <package name>` or `conda install <package name>`. In the following code snippet, first, we download NumPy, the fundamental package for scientific computing, via the `pip` command. Then, we will install PyTorch via the `conda` command. When installing PyTorch, you must provide a version for CUDA, a parallel computing platform and programming model that is used for general computing on GPUs. CUDA can speed up the DL model training by allowing GPUs to process the computation in parallel:

```
pip install numpy
conda install pytorch torchvision torchaudio \
cudatoolkit=<cuda version> -c pytorch -c nvidia
```

TensorFlow is another popular package for DL projects. Like PyTorch, TensorFlow provides different packages for each version of CUDA. The full list can be found online here: `https://www.tensorflow.org/install/source#gpu`. To get all libraries related to DL to work seamlessly together, there must be compatibility between the Python version, TensorFlow version, GCC compiler version, CUDA version, and Bazel build tool version, as shown in the following screenshot:

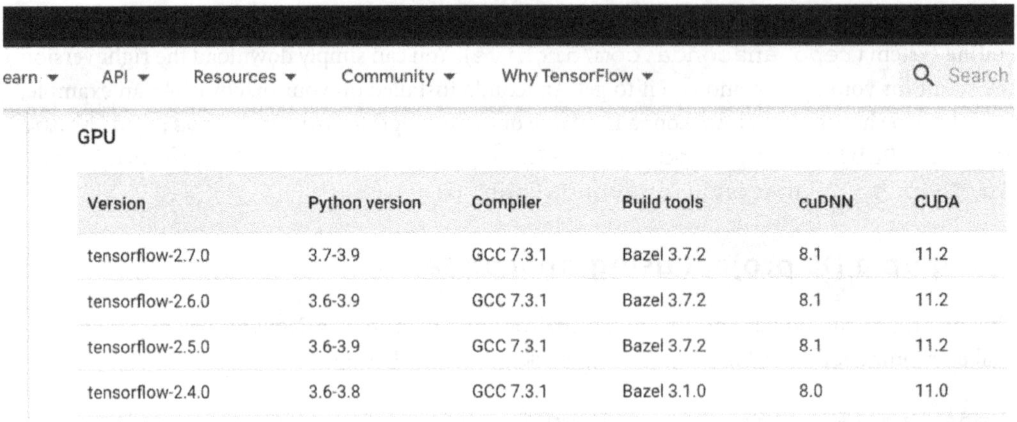

GPU

Version	Python version	Compiler	Build tools	cuDNN	CUDA
tensorflow-2.7.0	3.7-3.9	GCC 7.3.1	Bazel 3.7.2	8.1	11.2
tensorflow-2.6.0	3.6-3.9	GCC 7.3.1	Bazel 3.7.2	8.1	11.2
tensorflow-2.5.0	3.6-3.9	GCC 7.3.1	Bazel 3.7.2	8.1	11.2
tensorflow-2.4.0	3.6-3.8	GCC 7.3.1	Bazel 3.1.0	8.0	11.0

Figure 2.1 – Compatibility matrix for the TensorFlow, Python, GCC, Bazel, cuDNN, and CUDA versions

Going back to `pip` commands, instead of typing `install` commands repeatedly, you can generate a single text file that consists of the necessary packages and install all of them in a single command. To achieve this, you can provide the filename with the `--requirement` (`-r`) option to the `pip install` command, as follows:

```
pip install -r requirements.txt
```

Common packages required are listed in the CPU-only environments are listed in the sample `requirements.txt` file: `https://github.com/PacktPublishing/Production-Ready-Applied-Deep-Learning/blob/main/Chapter_2/anaconda/requirements.txt`. The main packages in the list are TensorFlow and PyTorch.

Now, let's look at some useful Anaconda commands. Just as `pip install` can be used with the `requirements.txt` file, you can also create an environment with a set of packages using a YAML file. In the following example, we are using an `env.yml` file to save the list of libraries from an existing environment. Later, `env.yml` can be used to create a new environment with the same packages, as presented in the following code snippet:

```
conda create -n env_1
conda activate env_1
# save environment to a file
conda env export > env.yml
# clone existing environment
conda create -n env_2 --clone env_1
# delete existing environment (env_1)
conda remove -n env_1 --all
```

```
# create environment (env_1) from the yaml file
conda env create -f env.yml
# using conda to install the libraries from requirements.txt
conda install --force-reinstall -y -q --name py37 -c conda-
forge --file requirements.txt
```

The following code snippet describes a sample YAML file generated from conda env export:

```
# env.yml
name: env_1
channels:
  - defaults
dependencies:
  - appnope=0.1.2=py39hecd8cb5_1001
  - ipykernel=6.4.1=py39hecd8cb5_1
  - ipython=7.29.0=py39h01d92e1_0
prefix: /Users/userA/opt/anaconda3/envs/new_env
```

The main components of this YAML file are the name of the environment (name), the source of the libraries (channels), and the list of libraries (dependencies).

Things to remember

a. Python is a standard language for data analysis due to its simple syntax

b. Python doesn't require explicit compilation

c. PIP is used for installing Python packages

d. Anaconda handles both Python package management and environment management

---

In the next section, we will explain how to collect data from various sources. Then, we will clean and preprocess the collected data for the following processes.

Data collection, data cleaning, and data preprocessing

In this section, we will introduce you to various tasks involved in the process of data collection. We will describe how to collect data from multiple sources and convert them into a generic form that data scientists can use regardless of the underlying task. This process can be broken down into a few parts: data collection, data cleaning, and data preprocessing. It is worth mentioning that task-specific transformation is considered feature extraction, which will be discussed in the following section.

Collecting data

First, we will introduce different data collection methods for composing initial datasets. Different techniques are necessary, depending on how the raw data is formatted. Most datasets are either available online as an HFML file or as a JSON object. Some data is stored in **Comma-Separated Values (CSV)** format, which can easily be loaded through the pandas library, a popular data analysis and manipulation tool. Hence, we will mainly focus on collecting HTML and JSON data and saving it in CSV format in this section. Additionally, we will present some popular dataset repositories.

Crawling web pages

Considered a fundamental component of the web, **HyperText Markup Language (HTML)** data is easily accessible and consists of diverse information. Consequently, the ability to crawl web pages can help you collect large amounts of interesting data. In this section, we will use BeautifulSoup, a Python-based web crawling library (`https://www.crummy.com/software/BeautifulSoup/`). As an example, we will demonstrate how to crawl Google Scholar pages and how to save the crawled data as a CSV file.

In this example, several functions of BeautifulSoup will be used to extract the author's first name, last name, email, research interests, citation count, h-index (high index), co-author, and paper titles. The following table shows the data that we wish to collect in this example:

Data of Interest	Description	Example
Author Name	First name of the author	William Eberle
Email	Verified email domain of the author	tntech.edu
Affiliation	Affiliated organization	Tennessee Technological University
Research Interests	Research interests of the author	Data Mining Anomaly Detection
Co-Authors	Co-authors who published papers with the author	G Stone D Talbert W Eberle

Table 2.1 – Data that can be collected from Google Scholar pages

Crawling a web page is a two-step process:

1. Utilize the requests library to get the HTML data in a `response` object.

2. Construct a `BeautifulSoup` object that parses the HTML tags in the `response` object.

These two steps can be summarized in the following code snippet:

```
# url points to the target google scholar page
response = requests.get(url)
html_soup = BeautifulSoup(response.text, 'html.parser')
```

The next step is to get the contents of interest from the BeautifulSoup object. *Table 2.2* summarizes common BeautifulSoup functions that let you extract the content of the interest from the parsed HTML data. Since our goal in this example is to store the collected data as a CSV file, we will simply generate a comma-separated string representation of the page and write it to a file. The complete implementation can be found at https://github.com/PacktPublishing/Production-Ready-Applied-Deep-Learning/blob/main/Chapter_2/google_scholar/google_scholar.py.

The following table provides the list of methods required for processing raw data from Google Scholar pages:

Method	Example	Explanation
find_all("<html_tag>", href=<True/False>)	find_all("a", href=True)	Find a specific tag with find_all that has the href attribute present in it.
find("<sub_string_pattern_to_find>")	find("/citations?user=")	The find method from the standard Python library is used to find the start index of the "/citations?user=" substring, which holds the unique string for the co-author.
select("<html_tag>#<class_name>")	select("div#gsc_prf_ivh")	select('div') is used to extract the text inside a div of the HTML page with a specific class name of the div. On the Google Scholar Page, the class name for the email is gsc_prf_ivh.
find_all("<html_tag>", class_="<class_name>")	find_all('href', class_="gsc_prf_ila")	find_all helps to get all href with a specified class name. On the Google Scholar page, you can use the gsc_prf_ila class name to extract the institute name associated with the author.
find_all("<image_tag>")	links = html_soup.select("img") for i in links: link = i["src"]	The find_all method helps get a list of image tags on a page that hold src attributes for the image URLs.

Table 2.2 – Possible feature extraction techniques

Next, we will learn about JSON, another popular raw data format.

Collecting JSON data

JSON is a language-independent format that stores data as key-value and/or key-array fields. Since most programming languages support key-value data structures (for example, a Dictionary in Python or a HashMap in Java), JSON is considered interchangeable (program independent). The following code snippet shows some sample JSON data:

```
{
    "first_name": "Ryan",
    "last_name": "Smith",
    "phone": [{"type": "home",
               "number": "111 222-3456"}],
    "pets": ["ceasor", "rocky"],
    "job_location": null
}
```

Have a look at the Awesome JSON Datasets GitHub repository (https://github.com/jdorfman/awesome-json-datasets), which contains a list of useful JSON data sources. Also, Public API's GitHub repository (https://github.com/public-apis/public-apis) consists of a list of web server endpoints where various JSON data can be retrieved. Additionally, we provide a script that collects JSON data from an endpoint and stores the necessary fields as a CSV file: https://github.com/PacktPublishing/Production-Ready-Applied-Deep-Learning/blob/main/Chapter_2/rest/get_rest_api_data.py. This example uses the Reddit dataset available at https://www.reddit.com/r/all.json.

Next, we will introduce popular public datasets in the fields of data science.

Popular dataset repositories

Besides web pages and JSON data, many public datasets can be used for various purposes. For example, you can get datasets from popular data hubs such as *Kaggle* (https://www.kaggle.com/datasets) or *MIT Data Hub* (https://datahub.csail.mit.edu/browse/public). These public datasets are often used for a wide range of activities by many research institutes as well as businesses. Data from varying domains such as healthcare, government, biology, and computer science are collected during research and donated to the repositories for the greater good. Like how these organizations manage and provide diverse datasets, community efforts exist for managing various public datasets: https://github.com/awesomedata/awesome-public-datasets.

Another popular source of datasets is data analytics libraries such as *sklearn*, *Keras*, and *TensorFlow*. The list of datasets provided by each library can be found at `https://scikit-learn.org/stable/datasets`, `https://keras.io/api/datasets/`, and `https://www.tensorflow.org/datasets`, respectively.

Finally, government organizations also provide many datasets to the public. For example, you can find interesting, curated datasets related to COVID in a data lake hosted by AWS: `https://dj2taa9i652rf.cloudfront.net`. From this list of datasets, you can easily download data on Moderna vaccination distribution among different states in CSV format by navigating to the `cdc-moderna-vaccine-distribution` page.

Now that you have collected an initial dataset, the next step is to clean it up.

Cleaning data

Data cleaning is the process of polishing raw data to keep the entries consistent. Common operations include filling up empty fields with default values, removing characters that are not alpha-numeric such as ? or !, removing stop words, and removing HTML tags such as `<p></p>`. Data cleaning also focuses on retaining relevant information from the collected data. For example, a user profile page may have a wide range of information, such as a biography, first name, email, and affiliations. During the data collection process, target information is extracted as-is so that it can be kept in the original HTML or JSON tags. In other words, the biographic information that's been collected might still have HTML tags for new lines (`<br>`) or bold (`<b></b>`), which do not add much value to the following analysis task. Throughout data cleaning, these unnecessary components should be dropped.

Before we discuss individual data cleaning operations, it would be nice to have some understanding of DataFrames, table-like data structures provided by the pandas library (`https://pandas.pydata.org/`). They have rows and columns, just like a SQL table or an Excel sheet. One of their fundamental functionalities is `pandas.read_csv`, which allows you to load a CSV file into a DataFrame, as demonstrated in the following code snippet. The `tabulate` library is a good pick for displaying the content on a terminal as the DataFrame structures the data in a table format.

The following code snippet shows how to read a CSV file and print the data using the `tabulate` library (in the proceeding example, `tabulate` will mimic the format of the Postgres psql CLI as we are using the `tablefmt="psql"` option):

```
import pandas as pd
from tabulate import tabulate
in_file = "../csv_data/data/cdc-moderna-covid-19-vaccine-
distribution-by-state.csv"
# read the CSV file and store the returned dataframe to a
variable "df_vacc"
```

```
df_vacc = pd.read_csv(in_file)
print(tabulate(df_vacc.head(5), headers="keys",
tablefmt="psql"))
```

The following screenshot shows the content of the DataFrame in the preceding code snippet after being displayed on a terminal using the `tabulate` library (you can view a similar output without the `tabulate` library by using `df_vacc.head(5)`). The following screenshot shows the allocation of vaccine doses for each jurisdiction:

	jurisdiction	week_of_allocations	_1st_dose_allocations	_2nd_dose_allocations
0	Connecticut	2021-06-21T00:00:00.000	41220	41220
1	Maine	2021-06-21T00:00:00.000	15800	15800
2	Massachusetts	2021-06-21T00:00:00.000	79500	79500

Figure 2.2 – Loading a CSV file using pandas and displaying the contents using tabulate

The first data cleaning operation we will discuss is filling in missing fields with default values.

Filling empty fields with default values

We will use the Google Scholar data we crawled earlier in this chapter to demonstrate how empty fields are filled with default values. After data inspection, you will find a few authors that have left their affiliations empty as they are unspecified:

author_name	email	affiliation
Ferrol Aderholdt	nvidia.com	NVIDIA
Khaled Rabieh	shsu.edu	nan

Figure 2.3 – The affiliation column contains missing values (nan)

The default value for each field differs based on the context and data type. For example, nine to six would be a typical default value for an operation hour, and an empty string would be a good choice for a missing middle name. The phrase, not applicable (N/A) is often used to explicitly indicate that the fields are empty. In our example, we will fill out the empty fields that contain `na` to indicate that the values were missing in the original web pages and not missed out due to errors throughout the collection process. The technique we will demonstrate in this example involves the `pandas` library; the DataFrame has a `fillna` method that fills the empty values in the specified value. The `fillna` method accepts a parameter value of `True` for updating the object in place without creating a copy of it.

The following code snippet explains how to fill the missing values in a DataFrame using the `fillna` method:

```
df = pd.read_csv(in_file)
# Fill out the empty "affiliation" with "na"
df.affiliation.fillna(value="na", inplace=True)
```

In the preceding code snippet, we loaded a CSV file into a DataFrame and set missing affiliation entries with na. This operation will be executed in place without creating an additional copy.

In the next section, we will describe how to remove stop words.

Removing stop words

Stop words are words that do not convey much value from an information retrieval perspective. Common English stop words include *its*, *and*, *the*, *for*, and *that*. As an example, entry of the research interest fields in Google Scholar data, we see *security and privacy preservation for wireless networks*. Words such as *and* and *for* are not useful when we interpret the meaning of this text. Therefore, removing these words is recommended in **natural language processing** (**NLP**) tasks. One of the most popular stop word removal features is provided by **Natural Language Toolkit** (**NLTK**), which is a suite of libraries and programs for symbolic and statistical NLP. The following are a few words that are considered as stop word tokens by the NLTK library:

```
['doesn', "doesn't", 'hadn', "hadn't", 'hasn', "hasn't", 'haven',
"haven't", 'isn', "isn't", 'ma', …]
```

Word tokenization is the process of breaking down a sentence into word tokens (word vectors). In general, it gets applied before stop word removal. The following code snippets demonstrate how to tokenize the `research_interest` fields of Google Scholar data and remove stop words:

```
import pandas as pd
import nltk
from nltk.stem import PorterStemmer
from nltk.tokenize import word_tokenize
import traceback
from nltk.corpus import stopwords
# download nltk corpuses
nltk.download('punkt')
nltk.download('stopwords')
# create a set of stop words
stop_words = set(stopwords.words('english'))
# read each line in dataframe (i.e., each line of input file)
for index, row in df.iterrows():
```

```
curr_research_interest = str(row['research_interest'])\
    .replace("##", " ")\
    .replace("_", " ")
# tokenize text data.
curr_res_int_tok = tokenize(curr_research_interest)
# remove stop words from the word tokens
curr_filtered_research = [w for w in curr_res_int_tok\
                            if not w.lower() in stop_words]
```

As you can see, we first download the stop words corpus for NLTK with stopwords. words('english') and remove word tokens that are not in the corpus. The full version is available at https://github.com/PacktPublishing/Production-Ready-Applied-Deep-Learning/blob/main/Chapter_2/data_preproessing/bag_of_words_tf_idf.py.

Like stop words, text that is not alpha-numeric does not add much value either. Therefore, we will explain how to remove them in the next section.

Removing text that is not alpha-numeric

Alpha-numeric characters are characters that are neither English alphabet characters nor numbers. For example, in the text "*Hi, How are you?*", there are two non-alpha-numeric characters: , and ?. As in the case of stop words, they can be dropped as they don't convey much information about the context. Once these characters are removed, the text will read *Hi How are you*.

To remove a set of specific characters, we can use **regular expressions (regex)**. Regex is a sequence of characters that represents a search pattern. The following *Table 2.3* shows a few important regex search patterns and explains what each means:

Regex Search Pattern	Explanation
[A-Za-z0-9]	Match any alphabet (lower or uppercase) or any numeric value. In short, match any alphanumeric character.
[^A-Za-z0-9]	^ indicates not. Match any character that is not alphanumeric.
+	Match the preceding pattern element one or more times. \W means match non-alphanumeric characters. \W+ indicates one or more non-alphanumeric characters.
?	Match the preceding pattern element zero or one time. Colou?r will match color since the character u preceding ? can occur zero times. This will also match colour since character u preceding ? occurs one time.
^	Matches the beginning of a line or string. ^a will match a string starting with apples and will not match ones that start with bags.

Table 2.3 – Key regex search patterns

You can find other useful patterns at https://docs.python.org/3/library/re.html.

Python provides a built-in regex library that supports finding and removing a set of texts that matches the given regular expression. The following code snippet shows how to remove characters that are not alphanumeric. The \W pattern matches any character that is not a word character. + after the pattern indicates that we would like to keep the preceding element one or more times. Putting them together, we will find one or more alphanumeric characters in the following code snippet:

```
def clean_text(in_str):
    clean_txt = re.sub(r'\W+', ' ', in_str)
    return clean_txt
# remove non alpha-numeric characters for feature "text"
text = clean_text(text)
```

As the last data cleaning operation, we will introduce how to drop newline characters efficiently.

Removing newlines

Finally, the collected text data may have unnecessary newline characters. In many cases, the trailing newline characters can be dropped without any harm, regardless of what the following tasks are. Such characters can be easily replaced by empty strings using Python's built-in replace functionality.

The following code snippet shows how to remove a newline in text:

```
# replace the new line in the given text with empty string.
text = input_text.replace("\n", "")
```

In the preceding code snippet, "abc\n" will turn into "abc".

The cleaned data often gets processed further so that the data represents the underlying data better. This process is called data preprocessing. We will take a deeper look into this process in the next section.

Data preprocessing

The goal of data preprocessing is to transform cleaned data into a generic form suitable for a wide range of data analytics tasks. There is not a clear distinction between data cleaning and data preprocessing. As a result, tasks such as replacing a set of texts or filling in missing values can be categorized as data cleaning, as well as data preprocessing. In this section, we will focus on techniques that were not covered in the previous section: normalization, converting text into lowercase, converting text into bag-of-words, and applying stemming to words.

Complete implementations of the following examples can be found at https://github.com/PacktPublishing/Production-Ready-Applied-Deep-Learning/tree/main/Chapter_2/data_preproessing.

Normalization

Sometimes, the values for a field might be represented differently, even though they mean the same thing. In the case of Google Scholar data, entries in research interests may be in different words, even though they refer to a similar domain. For example, data science, ML, and **artificial intelligence (AI)** refer to the same domain of AI in larger contexts. During the data preprocessing stage, we typically normalize them by converting ML and data science into AI, which represents the underlying information better. This helps the data science algorithms leverage the feature for the target task.

As demonstrated in the `normalize.py` script within the example repository, normalization for the preceding case can easily be achieved by keeping a dictionary that maps the expected value to the normalized value. In the following code snippet, `artificial_intelligence` will be the normalized value for the `data_science` and `machine_learning` features for `research_interests`:

```
# dictionary mapping the values are commonly used for
normalization
dict_norm = {"data_science": "artificial_intelligence",
    "machine_learning": "artificial_intelligence"}
# normalize.py
if curr in dict_norm:
    return dict_norm[curr]
else:
    return curr
```

The numeric values of a field also require normalization. For numeric values, normalization would be the process of rescaling each value into a specific range. In the following example, we are scaling each mean count of weekly vaccine distributions per state between 0 and 1. First, we calculate the mean counts for each state. Then, we compute the normalized mean count by dividing the mean counts by the maximum mean count:

```
# Step 1: calculate state-wise mean number for weekly corora
vaccine distribution
df = df_in.groupby("jurisdiction")["_1st_dose_allocations"]\
    .mean().to_frame("mean_vaccine_count").reset_index()
# Step 2: calculate normalized mean vaccine count
df["norm_vaccine_count"] = df["mean_vaccine_count"] / df["mean_
vaccine_count"].max()
```

The result of normalization can be seen in the following screenshot. The table in this screenshot consists of two columns – the mean vaccine count before normalization and after normalization:

```
+------------------+---------------------+----------------------+
| jurisdiction     | mean_vaccine_count  | norm_vaccine_count   |
+------------------+---------------------+----------------------+
| Alabama          |               52185 |             0.125401 |
| Alaska           |               10124 |            0.0243281 |
| American Samoa   |               312.5 |          0.000750942 |
+------------------+---------------------+----------------------+
```

Figure 2.4 – Normalized mean vaccine distribution per state

The next data preprocessing we will introduce is case conversion for text data.

Case conversion

In many cases, text data gets converted into lowercase or uppercase as a way of normalization. This brings some level of consistency, especially when the following tasks involve comparisons. In the stop words removal example, word tokens in the `curr_res_int_tok` variable are searched within the standard English stop words of the NLTK library. For the comparison to be successful, the case should be consistent. In the following code snippet, the tokens get converted into lowercase before the stop word search:

```
# word tokenize
curr_resh_int_tok = word_tokenize(curr_research_interest)
# remove stop words from the word tokens
curr_filtered_research = [w for w in curr_res_int_tok\
                    if not w.lower() in stop_words]
```

Another example can be found in `get_rest_api_data.py`, where we have collected and processed data from Reddit. In the following code snippet taken from the script, every text field gets converted into lowercase upon collection:

```
def convert_lowercase(in_str):
    return str(in_str).lower()
# convert string to lowercase
text = convert_lowercase(text)
```

Next, you will learn how stemming can improve the quality of the data.

Stemming

Stemming is the process of transforming a word into its root word. The benefit of stemming comes from keeping the words consistent if their underlying meaning is the same. For example, "*information*", "*informs*", and "*informed*" have the same root word – "*inform*". The following example shows how to utilize the NLTK library for stemming. The NLTK library offers a stemming feature based on *Porter stemming algorithm (Porter, Martin F. "An algorithm for suffix stripping." Program (1980))*:

```
from nltk.stem import PorterStemmer


# porter stemmer for stemming word tokens
ps = PorterStemmer()
word = "information"
stemmed_word = ps.stem(word) // "inform"
```

In the preceding code snippet, we instantiated `PosterStemmer` from the `nltk.stem` library and passed the text into the `stem` function.

Things to remember

a. Data comes in different formats such as JSON, CSV, HTML, and XML. There are many data collection tools available for each type of data.

b. Data cleaning is the process of polishing raw data to keep each entry consistent. Common operations include filling up empty features with default values, removing characters that are not alphanumeric, removing stop words, and removing unnecessary tags.

c. The goal of data preprocessing is to apply generic data augmentation to transform cleaned data into a form that is generic for any data analytic task.

d. The domain of data cleaning and data preprocessing overlaps, which means that some operations can be used for either process.

So far, we have discussed the generic processes for data preparation. Next, we will discuss the final process: feature extraction. Unlike the other processes we have covered, feature extraction involves task-specific operations. Let's take a closer look.

Extracting features from data

Feature extraction (**feature engineering**) is the process of transforming data into features that express the underlying information in a specific way for the target task. Data preprocessing applies generic techniques that are often necessary for most data analytics tasks. However, feature extraction requires you to exploit domain knowledge as it is specific to the task. In this section, we will introduce popular feature extraction techniques, including bag-of-words for text data, term frequency-inverse document frequency, converting color images into gray images, ordinal encoding, one-hot encoding, dimensionality reduction, and fuzzy match for comparing two strings.

Complete implementations of these examples can be found online at https://github.com/PacktPublishing/Production-Ready-Applied-Deep-Learning/tree/main/Chapter_2/data_preproessing.

First, we will start with the bag-of-words technique.

Converting text using bag-of-words

Bag-of-words (**BoW**) is a representation of a document that describes the occurrence of a set of words in the document (word frequency). BoW only considers the occurrence of words and ignores the order of the words or structures of words in the document. The sklearn library is one of the most widely used Python ML libraries that provides a simple interface for data preprocessing as well as model training. The `CountVectorizer` class from Sklearn helps to create BoW from text. The following code demonstrates how to use Sklearn features for BoW:

```
import pandas as pd
from sklearn.feature_extraction.text import CountVectorizer
document_1 = "This is a great place to do holiday shopping"
document_2 = "This is a good place to eat food"
document_3 = "One of the best place to relax is home"
# 1-gram (i.e., single word token used for BoW creation)
count_vector = CountVectorizer(ngram_range=(1, 1), stop_
words='english')
# transform the sentences
count_fit = count_vector.fit_transform([document_1, document_2,
document_3])
# create dataframe
df = pd.DataFrame(count_fit.toarray(), columns=count_vector.
get_feature_names_out())
print(tabulate(df, headers="keys", tablefmt="psql"))
```

The following screenshot summarizes the output of BoW in a table format:

	best	eat	food	good	great	holiday	home	place	relax	shopping
0	0	0	0	0	1	1	0	1	0	1
1	0	1	1	1	0	0	0	1	0	0
2	1	0	0	0	0	0	1	1	1	0

Figure 2.5 – Output of BoW on three sample documents

Next, we will introduce **term frequency-inverse document frequency (TF-IDF)** for text data.

Applying term frequency-inverse document frequency (TF-IDF) transformation

The problem with using word frequency is that the documents that have higher frequency will dominate the model or analysis. Hence, it is better to rescale the frequency based on how often a word occurs in all documents. Such scaling helps to penalize those highly frequent words (such as *the* and *have*) in a way that the numerical representation of the text expresses the context better.

Before introducing the formula for TF-IDF, we must define some notations. Let *n* be the total number of documents and *t* be a word (term). *df(t)* refers to the document frequency for word *t* (how many documents contain the word *t*), while *tf(t, d)* refers to the word *t* frequency in document *d* (how many times *t* appears in document *d*). With these definitions, we can define *idf(t)*, the inverse document frequency, as *log [n / df(t)] + 1*.

Overall, *tf-idf(t, d)*, *tf-idf* for word *t* and document *d* can be represented as *tf(t, d) * idf(t)*.

In the sample code script, bag_of_words_tf_idf.py, we are using research interest fields of Google Scholar data to calculate TF-IDF. Here, we utilize the TfidfVectorizer function of Sklearn. The fit_transform function takes in a set of documents and generates a TF-IDF-weighted document-term matrix. From this matrix, we can print out the top *N* research interests:

```
tfidf_vectorizer = TfidfVectorizer(use_idf=True)
# use the tf-idf instance to fit list of research_interest
tfidf = tfidf_vectorizer.fit_transform(research_interest_list)
# tfidf[0].T.todense() provides the tf-idf dense vector
# calculated for the research_interest
```

```
df = pd.DataFrame(tfidf[0].T.todense(), index=tfidf_vectorizer.
get_feature_names_out(), columns=["tf-idf"])
# sort the tf-idf calculated using 'sort_values' of dataframe.
df = df.sort_values('tf-idf', ascending=False)
# top 3 words with highest tf-idf
print(df.head(3))
```

In the preceding example, we create a `TfidfVectorizer` instance and trigger the `fit_transform` function using the list of research interest texts (`research_interest_list`). Then, we call the `todense` method on the output to obtain the dense representation of the resulting matrix. The matrix gets converted into a DataFrame and sorted to display the top entries. The following screenshot shows the output of `df.head(3)` – three words with the highest TF-IDF from research interest fields:

	tf-idf
anomaly	0.641387
detection	0.601448
mining	0.368282

Figure 2.6 – Three words with the highest TF-IDF from research interest fields

Next, you will learn how to process categorical data using one-hot encoding.

Creating one-hot encoding (one-of-k)

One-hot encoding (one-of-k) is the process of converting discrete values into a sequence of binary values. Let's start with a simple example, where a field can have categorical values of either cat or dog. The one-hot encoding will be represented by two bits, where one bit refers to cat and the other bit refers to dog. The bit in the encoding with a value of 1 means that the field has the corresponding value. So, 1 0 represents a cat, while 0 1 represents a dog:

breed	pet_type	dog	cat
Retrievers	dog	1	0
Maine Coon	cat	0	1
German Shepherd	dog	1	0

Table 2.4 – Converting categorical values in pet_type into one-hot encoding (dog and cat)

A demonstration of one-hot encoding can be found in `one_hot_encoding.py`. In the following code snippet, we are focusing on the core operations, which involve `OneHotEncoder` from Sklearn:

```
from sklearn.preprocessing import LabelEncoder
labelencoder = LabelEncoder()
encoded_data = labelencoder.fit_transform(df_research
['is_artifical_intelligent'])
```

The `is_artificial_intelligence` column used in the previous code snippet consists of two distinct values: "yes" and "no". The following screenshot summarizes the results of one-hot encoding:

```
+----+------------------------------------------+------+-------+
|    | is_artificial_intelligence        | no | yes |
|----+------------------------------------------+------+-------|
| 0 | yes                                    | 0 |   1 |
| 4 | no                                     | 1 |   0 |
+----+------------------------------------------+------+-------+
```

Figure 2.7 – One-hot encoding for the is_artificial_intelligence field

In the next section, we will introduce another type of encoding called ordinal encoding.

Creating ordinal encoding

Ordinal encoding is the process of converting a categorical value into a numerical value. In *Table 2.5*, there are two types for pets, dog and cat. Dogs are assigned a value of 1 and cats are assigned a value of 2:

breed	pet_type	ordinal_encoding
Retrievers	dog	1
Maine Coon	cat	2
German Shepherd	dog	1

Table 2.5 – The categorical values in pet_type field encoded as ordinal in ordinal_encoding

In the following code snippet, we are using the `LabelEncoder` class from Sklearn to transform research interest fields into numerical values. A complete example of ordinal encoding can be found in `ordinal_encoding.py`:

```
from sklearn.preprocessing import LabelEncoder
labelencoder = LabelEncoder()
```

```
encoded_data = labelencoder.fit_transform(df_research
['research_interest'])
```

The preceding code snippet is almost self-explanatory – we simply construct a `LabelEncoder` instance and pass the target column to the `fit_transform` method. The following screenshot shows the first three rows of the resulting DataFrame:

```
+------------------------------+------------------------------+
| research_interest            | encoded_research_interest    |
+------------------------------+------------------------------+
| data_mining                  |                          534 |
| anomaly_detection            |                          100 |
| artificial_intelligence      |                          128 |
```

Figure 2.8 – Results of ordinal encoding on research interest

Next, we will explain a technique for images: converting a colored image into a grayscale image.

Converting a colored image into a grayscale image

One of the most common techniques in a computer vision task is to convert a colored image into a grayscale image. *OpenCV* is a standard library for image processing (https://opencv.org/). In the following example, we are simply importing the OpenCV library (`import cv2`) and using the `cvtColor` function to convert a loaded image into grayscale:

```
image = cv2.imread('./images/tiger.jpg')
# filter to convert color tiger image to gray one
gray_image = cv2.cvtColor(image, cv2.COLOR_BGR2GRAY)
# write the gray image to a file
cv2.imwrite('./images/tiger_gray.jpg', gray_image)
```

When analyzing a large volume of data with multiple fields, you often find that reducing the number of dimensions is necessary. In the next section, we will look at the available options for this process.

Performing dimensionality reduction

In many cases, there are more features than what the task needs; not all features have useful information. In this case, you can use dimensionality reduction techniques such as **Principal Component Analysis (PCA)**, **Singular Value Decomposition (SVD)**, **Linear Discriminant Analysis (LDA)**, **t-SNE**, **UMAP**, and **ISOMAP** to name a few. Another option is to use DL. You can build a custom model for dimensionality reduction or use a pre-defined network structure such as **AutoEncoder**. In this section, we will describe PCA in detail as it is the most popular technique among the ones we mentioned.

Given a set of features, PCA identifies relationships among the features and generates a new set of variables that capture the differences in the samples in the most efficient way. These new variables are called principal components and are ranked in order of importance; while constructing the first principal component, it squeezes the unimportant variables and leaves them for the second principal component. Therefore, the first principal component is not correlated to the remaining variables. This process gets repeated to construct principal components of the following order.

If we were to describe the PCA process more formally, we can say that the process has two steps:

1. Constructs a covariance matrix that represents the correlations for every pair of features.

2. Generates a new set of features that captures different amounts of information by calculating the eigenvalues of the covariance matrix.

The new set of features is principal components. By sorting the corresponding eigenvalues from highest to lowest, you would get the most useful new feature at the top.

To understand the details, we will look at the Iris dataset (`https://archive.ics.uci.edu/ml/datasets/iris`). This dataset consists of three classes of Iris plant (setosa, versicolor, and virginica), along with four features (sepal width, sepal length, petal width, and petal length). In the following diagram, we plot each entry using the two new features constructed from PCA. Based on this diagram, we can easily conclude that we only need the top two principal components to distinguish the three classes:

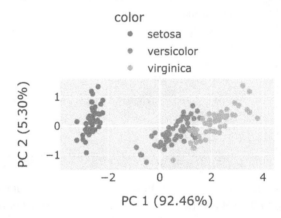

Figure 2.9 – The results of PCA on the Iris dataset

In the following example, we will use human resources data from Kaggle to demonstrate PCA. The initial set of data consists of multiple fields such as salary, whether there was a promotion within the last five years or not, and whether an employee left the company or not. Once principal components are constructed, they can be plotted using matplotlib:

```
import matplotlib.pyplot as plt
import numpy as np
```

```
import pandas as pd
from sklearn.decomposition import PCA
from sklearn.preprocessing import StandardScaler,
# read the HR data in csv format
df_features = pd.read_csv("./HR.csv")
# Step 1: Standardize features by removing the mean and scaling
to unit variance
scaler = StandardScaler()
# train = scaler.fit(X)
X_std = scaler.fit_transform(X)
# Step 2: Instantiate PCA & choose minimum number of
# components such that it covers 95% variance
pca = PCA(0.95).fit(X_std)
```

In the preceding code snippet, first, we loaded the data using the read_csv function of the pandas library, normalized the entries using StandardScaler from Sklearn, and applied PCA using Sklearn. The complete example can be found at pca.py.

As the last technique for feature extraction, we will explain how to effectively calculate a distance metric between two sequences.

Applying fuzzy matching to handle similarity between strings

Fuzzy matching (https://pypi.org/project/fuzzywuzzy/) uses a distance metric that measures the differences between two sequences and treats them equally if they can be considered similar. In this section, we will demonstrate how fuzzy matching can be applied using *Levenshtein Distance* (Levenshtein, Vladimir I. (February 1966). "Binary codes capable of correcting deletions, insertions, and reversals". Soviet Physics Doklady. 10 (8): 707–710. Bibcode: 1966SPhD...10..707L).

The most popular library for fuzzy string matching is fuzzywuzzy. The ratio function will provide the Levenshtein distance score, which we can use to decide whether we want to consider the two strings the same for the following process. The following code snippet describes the usage of the ratio function:

```
from fuzzywuzzy import fuzz
# compare strings using ratio method
fuzz.ratio("this is a test", "this is a test!") // 91
fuzz.ratio("this is a test!", "this is a test!") // 100
```

As shown in the preceding example, the ratio function will output a higher score if the two texts are more similar.

> **Things to remember**
>
> a. **Feature extraction** is the process of transforming data into features that express the underlying information better for the target task.
>
> b. BoW is a representation of a document based on the occurrence of the words. TF-IDF can express the context of a document better by penalizing highly frequent words.
>
> c. A colored image can be updated to a grayscale image using the OpenCV library.
>
> d. Categorical features can be represented numerically using ordinal encoding or one-hot encoding.
>
> e. When a dataset has too many features, dimensionality reduction can reduce the number of features that have the most information. PCA constructs new features while retaining most of the information.
>
> f. When evaluating the similarity between two texts, you can apply fuzzy matching drop.

Once the data has been transformed into a reasonable format, you will often need to visualize the data to understand its characteristics. In the next section, we will introduce popular libraries for data visualization.

Performing data visualization

When applying ML techniques to analyze a dataset, the first step must be understanding the available data because every algorithm has advantages that are closely related to the underlying data. The key aspects of data that data scientists need to understand include data formats, distributions, and relationships among the features. When the amount of data is small, necessary information can be collected by analyzing each entry manually. However, as the amount of data grows, visualization plays a critical role in understanding the data.

Many tools for data visualization are available in Python. **Matplotlib** and **Seaborn** are the most popular libraries for statistical data visualization. We will introduce these two libraries one by one in this section.

Performing basic visualizations using Matplotlib

In the following example, we will demonstrate how to generate bar charts and pie charts using Matplotlib. The data we use represents the weekly distribution of COVID vaccines. To use the `matplot` functionality, you must import the package first (`import matplotlib.pyplot as plt`). The `plt.bar` function takes the list of top 10 state names and a list of its mean distribution to generate a bar chart. Similarly, the `plt.pie` function is used to generate a pie chart from a dictionary. The `plt.figure` function resizes the plot size and allows users to draw multiple charts on the same canvas. The complete implementation can be found at `visualize_matplotlib.py`:

```
# PIE CHART PLOTTING
# colors for pie chart
colors = ['orange', 'green', 'cyan', 'skyblue', 'yellow',
'red', 'blue', 'white', 'black', 'pink']
# pie chart plot
plt.pie(list(dict_top10.values()), labels=dict_top10.keys(),
colors=colors, autopct='%2.1f%%', shadow=True, startangle=90)
# show the actual plot
plt.show()
# BAR CHART PLOTTING
x_states = dict_top10.keys()
y_vaccine_dist_1 = dict_top10.values()
fig = plt.figure(figsize=(12, 6))   # figure chart with size
ax = fig.add_subplot(111)
# bar values filling with x-axis/y-axis values
ax.bar(np.arange(len(x_states)), y_vaccine_dist_1, log=1)
plt.show()
```

The result of the preceding code is as follows:

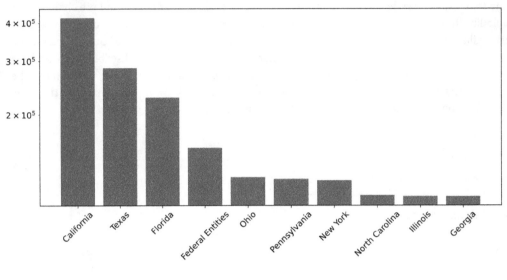

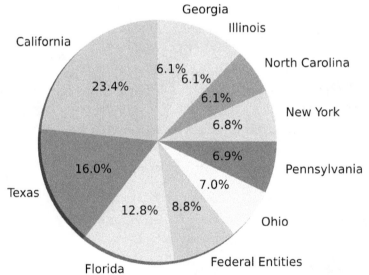

Figure 2.10 – Bar and pie charts generated using Matplotlib

In the next section, we will introduce Seaborn, another popular data visualization library.

Drawing statistical graphs using Seaborn

Seaborn is a library built on top of Matplotlib to provide a high-level interface for drawing statistical graphics that Matplotlib does not support. In this section, we will learn how to generate line graphs and histograms using Seaborn for the same dataset. First, we need to import the Seaborn library along with Matplotlib (import seaborn as sns). The sns.line_plot function is designed to accept a DataFrame and column names. Therefore, we must provide df_mean_sorted_top10, which contains the top 10 states of the highest mean values of vaccines distributed and two column names, state_names and count_vaccine, for the X and Y axes. To plot the histogram, you can use the sns.dist_plot function, which takes in a DataFrame with a column value for the Y axis. If we are to use the same mean values, it would be sns.displot(df_mean_sorted_top10['count_vaccine'], kde=False):

```
import seaborn as sns
# top 10 stats by largest mean
df_mean_sorted_top10 = ... # top 10 stats by largest mean
# LINE CHART PLOT
sns.lineplot(data=df_mean_sorted_top10, x="state_names",
y="count_vaccine")
# show the actual plot
plt.show()
# HISTOGRAM CHART PLOTTING
# plot histogram bars with top 10 states mean distribution
count of vaccine
sns.displot(df_mean_sorted_top10['count_vaccine'], kde=False)
plt.show()
```

The resulting graphs are shown in the following figure:

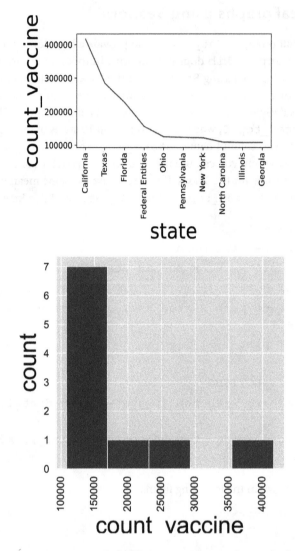

Figure 2.11 – Line graph and histogram generated using Seaborn

The complete implementation can be found in this book's GitHub repository (`visualize_seaborn.py`).

Many libraries can be used for enhanced visualizations: **pyROOT**, a data analysis framework from CERN that's commonly used for research projects (`https://root.cern/manual/python`), **Streamlit**, for easy web app creation (`https://streamlit.io`), **Plotly**, a free open source graphing library (`https://plotly.com`), and **Bokeh**, for interactive web visualizations (`https://docs.bokeh.org/en/latest`).

> **Things to remember**
>
> a. Visualizing data helps you analyze and understand data that is critical for selecting the right machine learning algorithm.
>
> b. Matplotlib and Seaborn are the most popular data visualization tools. Other tools include pyRoot, Streamlit, Plotly, and Bokeh.

The last section of this chapter will describe Docker, which allows you to achieve **operating system (OS)**-level virtualization for your project.

Introduction to Docker

In the previous section, *Setting up notebook environments*, you learned how to set up a virtual environment with various packages for DL using `conda` and `pip` commands. Furthermore, you know how to save an environment into a YAML file and recreate the same environment. However, projects based on virtual environments may not be sufficient when the environment needs to be replicated on multiple machines as there can be issues coming from non-obvious OS-level dependencies. In this situation, Docker would be a great solution. Using Docker, you can create a snapshot of your working environment, including the underlying version of your OS. Altogether, Docker allows you to separate your applications from your infrastructure so that you can deliver your software quickly. Installing Docker can be achieved by following the instructions at `https://www.docker.com/get-started`. In this book, we will use version 3.5.2.

In this section, we will introduce a Docker image, a representation of a virtual environment in the context of Docker, and explain how to create a Dockerfile for the target Docker image.

Introduction to Dockerfiles

Docker images are created by so-called Dockerfiles. Every Docker image has a base (or parent) image. For DL environments, a good choice for the base image would be an image developed for Linux Ubuntu OS – one of the following should be a good choice: *ubuntu:18.04* (`https://releases.ubuntu.com/18.04`) or *ubuntu:20.04* (`https://releases.ubuntu.com/20.04`). Along with an image for the underlying OS, there are images with specific packages already installed. For example, the simplest way to set up a TensorFlow-based environment is to download images with TensorFlow installed. A base image has been created by TensorFlow developers and can be easily downloaded by using `docker pull tensorflow/serving` command (`https://hub.docker.com/r/tensorflow/serving`). An environment with PyTorch is also available: `https://github.com/pytorch/serve/blob/master/docker/README.md`.

Next, you will learn how to build with a custom Docker image.

Building a custom Docker image

Creating a custom image is also straightforward. However, it involves many commands for which we will relegate the details to `https://github.com/PacktPublishing/Production-Ready-Applied-Deep-Learning/tree/main/Chapter_2/dockerfiles`. Once you have built the Docker image, you can instantiate it with something known as a Docker container. A Docker container is a standalone executable package of software that includes everything that you need to run the target application. By following the instructions in the `README.md` file, you can create the Docker image, which will run a containerized Jupyter notebook service with the standard libraries we described in this chapter.

> **Things to remember**
>
> a. Docker creates a snapshot of your working environment, including the underlying OS. The created image can be used to recreate the same environment on different machines.
>
> b. Docker helps you separate your environment from infrastructure. This allows you to move your applications to different cloud service providers (such as AWS or Google Cloud) with minimal effort.

At this point, you should be able to create a Docker image for your DL project. By instantiating the Docker image, you should be able to collect the data you need and process it as needed on your local machine or various cloud service providers.

Summary

In this chapter, we described how to prepare a dataset for data analytics tasks. The first key point was how to achieve environment virtualization using Anaconda and Docker, along with Python package management using `pip`.

The data preparation process can be broken down into four steps: data collection, data cleaning, data preprocessing, and feature extraction. First, we have introduced various tools available for data collection that support different data types. Once the data has been collected, it is cleaned and preprocessed so that it can be transformed into a generic form. Depending on the target task, we often apply various feature extraction techniques that are task-specific. In addition, we have introduced many tools for data visualization that can help you understand the characteristics of the prepared data.

Now that we have learned how to prepare our data for analytics tasks, in the next chapter, we will explain DL model development. We will introduce the basic concepts and how to use the two most popular DL frameworks: TensorFlow and PyTorch.

3

Developing a Powerful Deep Learning Model

In this chapter, we will describe how to design and train a **deep learning** (**DL**) model. Within the notebook context described in the previous chapter, data scientists investigate various network designs and model training settings to generate a working model for the given task. The main topics of this chapter include the theory behind DL and how to train a model using the most popular DL frameworks: **PyTorch** and **TensorFlow** (**TF**). At the end of the chapter, we will decompose the **StyleGAN** implementation, a popular DL model for image generation, to explain how to construct a complex model using the components that we have introduced in this chapter.

In this chapter, we're going to cover the following main topics:

- Going through the basic theory of DL
- Understanding the components of DL frameworks
- Implementing and training a model in PyTorch
- Implementing and training a model in TF
- Decomposing a complex, state-of-the-art model implementation

Technical requirements

You can download the supplemental material of this chapter from the following GitHub link: https://github.com/PacktPublishing/Production-Ready-Applied-Deep-Learning/tree/main/Chapter_3.

The samples in this chapter can be executed from any Python environment with the necessary packages installed. You can use the sample environment introduced in the last chapter: https://github.com/PacktPublishing/Production-Ready-Applied-Deep-Learning/tree/main/Chapter_2/dockerfiles.

Going through the basic theory of DL

As briefly described in *Chapter 1, Effective Planning of Deep-Learning-Driven Projects*, DL is a **machine learning (ML)** technique based on **artificial neural networks (ANNs)**. In this section, our goal is to explain how ANNs work without going too deep into the math.

How does DL work?

An ANN is basically a set of connected neurons. As shown in *Figure 3.1*, neurons from an ANN and neurons from our brain behave in a similar way. Each connection in an ANN consists of a tunable parameter called the **weight**. When there is a connection from neuron A to neuron B, the output of neuron A gets multiplied by the weight of the connection; the weighted value becomes the input of neuron B. **Bias** is another tunable parameter within a neuron; a neuron sums up all the inputs and adds the bias. The last operation is an activation function that maps the computed value into a different range. The value in the new range is the output of the neuron, which gets passed to other neurons based on the connections.

Throughout the research, it has been found that groups of neurons captures different patterns based on their organization. Some of the powerful organizations are standardized as **layers** and have become the main building block for an ANN, providing a layer of abstraction on top of the complicated interactions among the neurons.

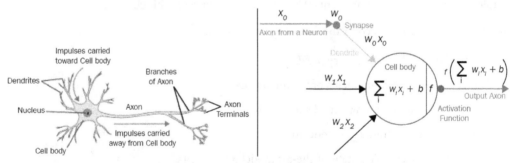

Drawing of a Biological Neuron (Left) and its Mathematical Model (Right).

Figure 3.1 – A comparison of a biological neuron and a mathematical model of an ANN neuron

As described in the preceding diagram, operations in DL are based on numerical values. Therefore, the input data for a network must be converted into a numerical value. For example, a **Red, Green, and Blue (RGB)** color code is a standard way of representing an image using numerical values. In the case of text data, word embeddings are often used. Similarly, the output of a network will be a set of numerical values. The interpretation of these values can vary based on the task and the definition.

DL model training

Overall, training an ANN is a process of finding a set of weights, biases, and activation functions that enable the network to extract meaningful patterns from the data. Now, the next question would be the following: *how do we find the right set of parameters?* Many researchers have tried to solve this problem using various techniques. Out of all the trials, the most effective algorithm discovered is an optimization algorithm called **gradient descent**, an iterative process that finds the local or global minimum.

When training a DL model, we need to define a function that quantizes the difference between predictions and ground-truth labels as a numeric value called a **loss**. With a loss function clearly defined, we iteratively generate intermediate predictions, compute loss values, and update model parameters in the direction toward the minimum loss.

Given that the goal of optimization is to find the minimum loss, model parameters need to be updated based on the **train set** samples in the opposite direction of the gradient (see *Figure 3.2*). To compute the gradients, the network keeps track of the intermediate values computed during the prediction pass (**forward propagation**). Then, starting from the last layer, it computes the gradients for each parameter exploiting the chain rule (**backward propagation**). Interestingly, model performance and training time can differ a lot based on how the parameters get updated in each iteration. The different parameter updating rules are captured within the concept of optimizers. One of the main tasks in DL is to select the type of optimizer that produces the model with the best performance.

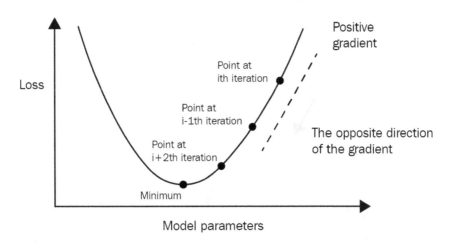

Figure 3.2 – With gradient descent, model parameters will be updated
in the opposite direction of the gradient at every iteration

However, there is one caveat to this process. If the model is trained to achieve the best performance for the train set specifically, the performance on unseen data can possibly deteriorate. This is called **overfitting**; the model is trained specifically for the data it has seen before and fails to make correct predictions on new data. On the other hand, a shortage of training can lead to **underfitting**, a situation in which the model fails to capture the underlying pattern of the train set. To prevent these issues, a portion of the train set is put aside for evaluating the trained model throughout the training: the **validation set**. Overall, training for DL involves a process of updating the model parameters based on the train set but selecting the model that performs the best on the validation set. The last type of dataset, the **test set**, represents what the model would interact with once it is deployed. The test set may or may not be available at the time of model training. The purpose of the test set is to understand how the trained model would perform in production. To further understand the overall training logic, we can look at *Figure 3.3*:

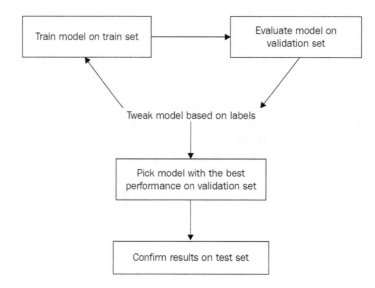

Figure 3.3 – The steps for training a DL model

The figure clearly describes what steps there are within the iterative process and what role each type of dataset plays in the scene.

Things to remember

a. Training an ANN is a process of finding a set of weights, biases, and activation functions that enable the network to extract meaningful patterns from the data.

b. There are three types of datasets in the training flow. The model parameters are updated using the train set, and the one that produces the best performance on the validation set is selected. The test set reflects the data distribution that the trained model would interact with upon deployment.

Next, we will look at DL frameworks that are designed to help us with model training.

Components of DL frameworks

Since the configuration of model training follows the same process regardless of the underlying tasks, many engineers and researchers have put together the common building blocks into frameworks. Most of the frameworks simplify DL model development by keeping data loading logic and model definitions independent from the training logic.

The data loading logic

Data loading logic includes everything from loading the raw data in memory to preparing each sample for training and evaluation. In many cases, data for the train set, validation set, and test set are stored in separate locations, so that each of them requires a distinct loading and preparation logic. The standard frameworks keep these logics separate from the other building blocks so that the model can be trained using different datasets in a dynamic way with minimal changes on the model side. Furthermore, the frameworks have standardized the way that these logics are defined to improve reusability and readability.

The model definition

Another building block, **model definition**, refers to the ANN architecture itself and corresponding forward and backward propagation logics. Even though building up a model using arithmetic operations is an option, the standard frameworks provide common layer definitions that users can put together to build up a complex model. Therefore, users are responsible for instantiating the necessary network components, connecting the components, and defining how the model should behave for training and inference.

In the following two sections, *Implementing and training a model in PyTorch* and *Implementing and training a model in TF*, we will introduce how to instantiate the popular layers in PyTorch and TF, respectively: dense (linear), pooling, normalization, dropout, convolution, and recurrent layers.

Model training logic

Lastly, we need to combine the two components and define the details of the training logic. This wrapper component must clearly describe the essential pieces of the model training, such as loss function, learning rate, optimizer, epochs, iterations, and batch size.

Loss functions can be classified into two major categories based on the type of learning task: **classification loss** and **regression loss**. The major difference between the two categories comes from the output format; the output of the classification task is categorical, while the output of the regression task is a continuous value. Out of the different losses, we will mainly discuss **Mean Square Error (MSE) loss** and **Mean Absolute Error (MAE) loss** for regression loss, and **Cross-Entropy (CE) loss** and **Binary Cross-Entropy (BCE) loss** for classification loss.

The **learning rate** (**LR**) defines the size of a step that gradient descent takes in the direction of the local minimum. Selecting the LR rate will help the process to converge faster, but if it's too high or low, the convergence will not be guaranteed (see *Figure 3.4*):

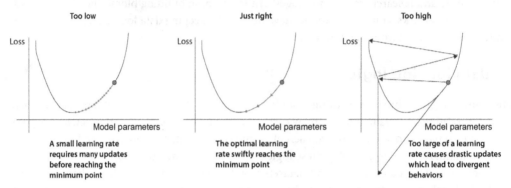

Figure 3.4 – The impact of the LR within gradient descent

Speaking of **optimizers**, we focus on the two main optimizers: **Stochastic Gradient Descent** (**SGD**), a basic optimizer with a fixed LR, and **Adaptive Moment Estimation** (**Adam**), an optimizer based on an adaptive LR that works the best in most scenarios. If you are interested in learning about different optimizers and the mathematics behind them, we recommend reading a survey paper by Choi et al (`https://arxiv.org/pdf/1910.05446.pdf`).

A single **epoch** indicates that every sample in the train set has been passed forward and backward through the network and that the network parameters have been updated. In many cases, the number of samples in the train set is way too huge to be passed through in one queue, so it gets divided into **mini-batches**. The **batch size** refers to the number of samples in a single mini-batch. Given that a set of mini-batches makes up the whole dataset, the number of iterations refers to the number of gradient update events (more precisely, the number of mini-batches) that model needs to interact with every sample. For example, if a mini-batch has 100 samples and there are 1,000 samples in total, it will require 10 iterations to complete one epoch. Selecting the right number of epochs is not an easy task. It changes depending on the other training parameters such as LR and batch size. Therefore, it often requires a trial-and-error process, keeping underfitting and overfitting in mind.

Things to remember

a. The components of model training can be broken down into data loading logic, model definition, and model training logic.

b. Data loading logic includes everything from loading raw data in the memory to preparing each sample for training and evaluation.

c. Model definition refers to the definition of the network architecture and its forward and backward propagation logics.

d. Model training logic handles the actual training by putting data loading logic and model definition together.

Out of the various frameworks available, we will discuss the two most popular in this book: **TF** and **PyTorch**. **Keras** running on TF has gained popularity in today, while PyTorch is heavily used for research with its exceptional flexibility and simplicity.

Implementing and training a model in PyTorch

PyTorch is a Python library for Torch, a ML package for Lua. The main features of PyTorch include **graphics processing unit**- (**GPU**-) accelerated matrix calculation and automatic differentiation for building and training neural networks. Creating the computation graph dynamically as the code gets executed, PyTorch is gaining popularity for its flexibility and ease of use, as well as its efficiency in model training.

Built on top of PyTorch, **PyTorch Lightning** (**PL**) provides another layer of abstraction, hiding many boilerplate codes. The new framework pays more attention to researchers by decoupling research-related components of PyTorch from the engineering-related components. PL codes are typically more scalable and easier to read than PyTorch codes. Even though the code snippets in this book put more emphasis on PL, PyTorch and PL share a lot of functionalities, so most components are interchangeable. If you are willing to dig into the details, we recommend the official site, `https://pytorch.org`.

There are other extensions of PyTorch available on the market:

- Skorch (`https://github.com/skorch-dev/skorch`) – A scikit-learn compatible neural network library that wraps PyTorch
- Catalyst (`https://github.com/catalyst-team/catalyst`) – A PyTorch framework specialized for reproducibility, rapid experimentation, and codebase reuse
- Fastai (`https://github.com/fastai/fastai`) – A library that standardizes not only high-level components for practitioners but also delivers low-level components for researchers
- PyTorch Ignite (`https://pytorch.org/ignite/`) – A library designed to help with training and evaluation for practitioners

We will not cover these libraries in this book, but you may find them helpful if you are new to this field.

Now, let's dive into PyTorch and PL.

PyTorch data loading logic

For readability and modularity, PyTorch and PL exploit a class called `Dataset` for data management and another class, `DataLoader`, for accessing samples iteratively.

While the `Dataset` class handles fetching individual samples, model training takes in the input data in batches and requires reshuffling to reduce model overfitting. `DataLoader` abstracts this complexity for users by providing a simple API. Furthermore, it exploits Python's multiprocessing features behind the scenes to speed up data retrieval.

The two core functions that must be implemented by the child class of `Dataset` are `__len__` and `__getitem__`. As described in the following class outline, `__len__` should return the total number of samples and `__getitem__` should return a sample for the given index:

```
from torch.utils.data import Dataset
class SampleDataset(Dataset):
    def __len__(self):
        """return number of samples"""
    def __getitem__(self, index):
        """loads and returns a sample from the dataset at the
given index"""
```

PL's `LightningDataModule` encapsulates all the steps needed to process data. The key components include downloading and cleaning data, preprocessing each sample, and wrapping each type of dataset inside `DataLoader`. The following code snippet describes how to create a `LightningDataModule` class. The class has the `prepare_data` function for downloading and preprocessing the data, as well as three functions for instantiating `DataLoader` for each type of dataset, `train_dataloader`, `val_dataloader`, and `test_dataloader`:

```
from torch.utils.data import DataLoader
from pytorch_lightning.core.lightning import
LightningDataModule
class SampleDataModule(LightningDataModule):
    def prepare_data(self):
        """download and preprocess the data; triggered only on
single GPU"""
        ...
    def setup(self):
        """define necessary components for data loading on each
GPU"""
        ...
    def train_dataloader(self):
        """define train data loader"""
        return data.DataLoader(
            self.train_dataset,
                batch_size=self.batch_size,
                shuffle=True)
    def val_dataloader(self):
```

```
        """define validation data loader"""
        return data.DataLoader(
            self.validation_dataset,
            batch_size=self.batch_size,
            shuffle=False)
    def test_dataloader(self):
        """define test data loader"""
        return data.DataLoader(
            self.test_dataset,
            batch_size=self.batch_size,
            shuffle=False)
```

The official documentation for `LightningDataModule` can be found at `https://pytorch-lightning.readthedocs.io/en/stable/extensions/datamodules.html`.

PyTorch model definition

The key benefit of PL comes from `LightningModule`, which simplifies the organization of complex PyTorch codes into six sections:

- Computation (`__init__`)
- The train loop (`training_step`)
- The validation loop (`validation_step`)
- The test loop (`test_step`)
- The prediction loop (`predict_step`)
- Optimizers and LR scheduler (`configure_optimizers`)

The model architecture is part of the computation section. Necessary layers are instantiated inside the `__init__` method, and computational logics are defined in the `forward` method. In the following code snippet, three linear layers are registered to the `LightningModule` module inside the `__init__` method, and the relationships between them are defined inside the `forward` method:

```python
from pytorch_lightning import LightningModule
from torch import nn
class SampleModel(LightningModule):
    def __init__(self):
        """instantiate necessary layers"""
```

```
        self.individual_layer_1 = nn.Linear(..., ...)
        self.individual_layer_2 = nn.Linear(..., ...)
        self.individual_layer_3 = nn.Linear(..., ...)
    def forward(self, input):
        """define forward propagation logic"""
        output_1 = self.individual_layer_1(input)
        output_2 = self.individual_layer_2(output_1)
        final_output = self.individual_layer_3(output_2)
        return final_output
```

Another way of defining a network is to use `torch.nn.Sequential`, as shown in the following code. With this module, a set of layers can be grouped together, and output chaining is automatically achieved:

```
class SampleModel(LightningModule):
    def __init__(self):
        """instantiate necessary layers"""
        self.multiple_layers = nn.Sequential(
        nn.Linear(     ,     ),
        nn.Linear(     ,     ),
        nn.Linear(     ,     ))
    def forward(self, input):
        """define forward propagation logic"""
        final_output = self.multiple_layers(input)
        return final_output
```

In the preceding code, the three linear layers are grouped together and stored as a single instance variable, `self.multiple_layers`. In the `forward` method, we simply trigger `self.multiple_layers` with the input tensor to pass the tensor through each layer one by one.

The following section is designed to introduce popular layer implementations.

PyTorch DL layers

One of the major benefits of DL frameworks comes from various layer definitions: gradient calculation logics are already part of the layer definitions, so you can focus on finding the best model architecture for your task. In this section, we will learn about layers that are commonly used across projects. Please refer to the official documentation (`https://pytorch.org/docs/stable/nn.html`) if the layer that you are interested in is not covered in this section.

PyTorch dense (linear) layer

The first type of layer is `torch.nn.Linear`. As the name suggests, it applies a linear transformation to the input tensor. The two main parameters of the function are `in_features` and `out_features`, which define the input and output tensor dimensions, respectively:

```
linear_layer = torch.nn.Linear(
            in_features,     # Size of each input sample
            out_features,    # Size of each output sample)
# N = batch size
# * = any number of additional dimensions
input_tensor = torch.rand(N, *, in_features)
output_tensor = linear_layer(input_tensor) # (N, *, out_
features)
```

The layer implementation from the `torch.nn` module already has the `forward` function defined, so that you can use the layer variable as if it were a function to trigger forward propagation.

PyTorch pooling layers

Pooling layers are commonly used for downsampling a tensor. The two most popular types are maximum pooling and average pooling. The key parameters for these layers are `kernel_size` and `stride`, which define the size of the window and how it moves for each pooling operation.

The maximum pooling layer downsamples the input tensor by selecting the largest value for each window:

```
# 2D max pooling
max_pool_layer = torch.nn.MaxPool2d(
    kernel_size,           # the size of the window to take a max
over
    stride=None,           # the stride of the window. Default
value is kernel_size
    padding=0,             # implicit zero padding to be added on
both sides
    dilation=1,            # a parameter that controls the stride
of elements in the window)
# N = batch size
# C = number of channels
# H = height of input planes in pixels
# W = width of input planes in pixels
input_tensor = torch.rand(N, C, H, W)
```

```
output_tensor = max_pool_layer(input_tensor) # (N, C, H_out,
W_out)
```

On the other hand, the average pooling layer downsamples the input tensor by computing an average value for each window:

```
# 2D average pooling
avg_pool_layer = torch.nn.AvgPool2d(
    kernel_size,          # the size of the window to take a max
over
    stride=None,          # the stride of the window. Default
value is kernel_size
    padding=0,            # implicit zero padding to be added on
both sides)
# N = batch size
# C = number of channels
# H = height of input planes in pixels
# W = width of input planes in pixels
input_tensor = torch.rand(N, C, H, W)
output_tensor = avg_pool_layer(input_tensor) # (N, C, H_out,
W_out)
```

You can find the other types of pooling layers at https://pytorch.org/docs/stable/nn.html#pooling-layers.

PyTorch normalization layers

Commonly used in data processing, the purpose of normalization is to scale numerical data to a common scale without distorting the distribution. In the case of DL, normalization layers are used to train the network with greater numerical stability (https://pytorch.org/docs/stable/nn.html#normalization-layers).

The most popular normalization layer is the batch normalization layer, which scales a set of values in a mini-batch. In the following code snippet, we introduce torch.nn.BatchNorm2d, a batch normalization layer designed for a mini-batch of 2D tensors with an additional channel dimension:

```
batch_norm_layer = torch.nn.BatchNorm2d(
    num_features,         # Number of channels in the input image
    eps=1e-05,            # A value added to the denominator for
numerical stability
```

```
    momentum=0.1,        # The value used for the running_mean and
running_var computation
    affine=True,         # a boolean value that when set to True,
this module has learnable affine parameters)
# N = batch size
# C = number of channels
# H = height of input planes in pixels
# W = width of input planes in pixels
input_tensor = torch.rand(N, C, H, W)
output_tensor = batch_norm_layer(input_tensor) # same shape as
input (N, C, H, W)
```

Out of the various parameters, the main one that you should be aware of is num_features, which indicates the number of channels. The input to the layer is a 4D tensor, where each index indicates the batch size (N), number of channels (C), the height of the image (H), and the width of the image (W).

PyTorch dropout layer

The dropout layer helps the model to extract generic features by randomly setting a set of values to zero. This operation prevents the model from overfitting to the train set. Having said that, the dropout layer implementation of PyTorch mainly operates over a single parameter, p, which controls the probability of an element being zeroed:

```
drop_out_layer = torch.nn.Dropout2d(
    p=0.5,  # probability of an element to be zeroed )
# N = batch size
# C = number of channels
# H = height of input planes in pixels
# W = width of input planes in pixels
input_tensor = torch.rand(N, C, H, W)
output_tensor = drop_out_layer(input_tensor) # same shape as
input (N, C, H, W)
```

In this example, we are dropping 50% of the elements (p=0.5). Similar to the batch normalization layer, the input tensor for torch.nn.Dropout2d has a size of N, C, H, W.

PyTorch convolution layers

Specialized for image processing, the convolutional layer is designed to apply convolution operations over the input tensor using a sliding window technique. In the case of image processing, where intermediate data is represented as 4D tensors of size N, C, H, W, torch.nn.Conv2d is the standard choice:

```
conv_layer = torch.nn.Conv2d(
    in_channels,            # Number of channels in the input image
    out_channels,           # Number of channels produced by the
convolution
    kernel_size,            # Size of the convolving kernel
    stride=1,               # Stride of the convolution
    padding=0,              # Padding added to all four sides of
the input.
    dilation=1,             # Spacing between kernel elements)
# N = batch size
# C = number of channels
# H = height of input planes in pixels
# W = width of input planes in pixels
input_tensor = torch.rand(N, C_in, H, W)
output_tensor = conv_layer(input_tensor) # (N, C_out, H_out,
W_out)
```

The first parameter of the torch.nn.Conv2d class, in_channels, indicates the number of channels in the input tensor. The second parameter, out_channels, indicates the number of channels in the output tensor, which is equal to the number of filters. The other parameters, kernel_size, stride, and padding, determine how the convolution operations are carried out for the layer.

PyTorch recurrent layers

Recurrent layers are designed for sequential data. Among the various types of recurrent layers, we will cover torch.nn.RNN in this section, which applies a multi-layer Elman **recurrent neural network (RNN)** to the given sequence (https://onlinelibrary.wiley.com/doi/abs/10.1207/s15516709cog1402_1). If you would like to try different recurrent layers, you can refer to the official documentation: https://pytorch.org/docs/stable/nn.html#recurrent-layers:

```
# multi-layer Elman RNN with tanh or ReLU non-linearity to an
input sequence.
rnn = torch.nn.RNN(
    input_size,                         # The number of expected
features in the input x
    hidden_size,                        # The number of features in
the hidden state h
```

```
    num_layers = 1,                    # Number of recurrent layers
    nonlinearity="tanh",               # The non-linearity to use.
Can be either 'tanh' or 'relu'
    bias=True,                         # If False, then the layer
does not use bias weights
    batch_first=False,                 # If True, then the input
and output tensors are provided
```

```
# as (batch, seq, feature) instead of (seq, batch, feature)
                dropout=0,
# If non-zero, introduces a Dropout layer on the outputs of
each RNN layer
```

```
# except the last layer, with dropout probability equal to
dropout
    bidirectional=False,               # If True, becomes a
bidirectional RNN)
# N = batch size
# L = sequence length
# D = 2 if bidirectionally, otherwise 1
# H_in = input_size
# H_out = hidden_size
rnn = nn.RNN(H_in, H_out, num_layers)
input_tensor = torch.randn(L, N, H_in)
# H_0 = tensor containing the initial hidden state for each
element in the batch
h0 = torch.randn(D * num_layers, N, H_out)
# output_tensor (L, N, D * H_out)
# hn (D * num_layers, N, H_out)
output_tensor, hn = rnn(input_tensor, h0)
```

The three key parameters of torch.nn.RNN are input_size, hidden_size, and num_layers. They refer to the number of expected features in the input tensor, the number of features in the hidden state, and the number of recurrent layers to use, respectively. To trigger forward propagation, you need to pass two things, an input tensor and a tensor containing the initial hidden state.

PyTorch model training

In this section, we describe the model training component of PL. As shown in the following code block, `LightningModule` is the base class that you must inherit for this component. Its `configure_optimizers` function is used to define the optimizer for training. Then, the actual training logic is defined within the `training_step` function:

```python
class SampleModel(LightningModule):
    def configure_optimizers(self):
        """Define optimizer to use"""
        return torch.optim.Adam(self.parameters(), lr=0.02)

    def training_step(self, batch, batch_idx):
        """Define single training iteration"""
        x, y = batch
        y_hat = self(x)
        loss = F.cross_entropy(y_hat, y)
        return loss
```

Validation, prediction, and the test loop have similar function definitions; a batch gets fed into the network to compute the necessary predictions and loss values. The collected data can also be stored and displayed using PL's built-in logging system. For details, please refer to the official documentation (https://pytorch-lightning.readthedocs.io/en/latest/common/lightning_module.html):

```python
    def validation_step(self, batch, batch_idx):
        """Define single validation iteration"""
        loss, acc = self._shared_eval_step(batch, batch_idx)
        metrics = {"val_acc": acc, "val_loss": loss}
        self.log_dict(metrics)
        return metrics

    def test_step(self, batch, batch_idx):
        """Define single test iteration"""
        loss, acc = self._shared_eval_step(batch, batch_idx)
        metrics = {"test_acc": acc, "test_loss": loss}
        self.log_dict(metrics)
        return metrics
```

```
def _shared_eval_step(self, batch, batch_idx):
    x, y = batch
    outputs = self(x)
    loss = self.criterion(outputs, targets)
    acc = accuracy(outputs.round(), targets.int())
    return loss, acc


def predict_step(self, batch, batch_idx, dataloader_idx=0):
    """Compute prediction for the given batch of data"""
    x, y = batch
    y_hat = self(x)
    return y_hat
```

Under the hood, LightningModule executes the following set of simplified PyTorch codes:

```
model.train()
torch.set_grad_enabled(True)
outs = []
for batch_idx, batch in enumerate(train_dataloader):
    loss = training_step(batch, batch_idx)
    outs.append(loss.detach())
    # clear gradients
    optimizer.zero_grad()
    # backward
    loss.backward()
    # update parameters
    optimizer.step()
    if validate_at_some_point
        model.eval()
        for val_batch_idx, val_batch in enumerate(val_
dataloader):
            val_out = model.validation_step(val_batch,
val_batch_idx)
        model.train()
```

Putting `LightningDataModule` and `LightningModule` together, the training and inference on the test set can be simply achieved as follows:

```
from pytorch_lightning import Trainer
data_module = SampleDataModule()
trainer = Trainer(max_epochs=num_epochs)
model = SampleModel()
trainer.fit(model, data_module)
result = trainer.test()
```

By now, you should've learned what you need to implement to set up a model training using PyTorch. The following two sections are dedicated to loss functions and optimizers, the two major components of model training.

PyTorch loss functions

First, we will look at the different loss functions available in PL. The loss functions in this sections can be found from the `torch.nn` module.

PyTorch MSE / L2 loss function

MSE loss function can be created using `torch.nn.MSELoss`. However, this calculates the square error component only and exploits the `reduction` parameter to provide variations. When `reduction` is None, the calculated value is returned as is. On the other hand, when it is set to `sum`, the outputs will be summed up. To obtain the exact MSE loss, the reduction must be set to `mean`, as shown in the following code snippet:

```
loss = nn.MSELoss(reduction='mean')
input = torch.randn(3, 5, requires_grad=True)
target = torch.randn(3, 5)
output = loss(input, target)
```

Next, let's have a look at MAE loss.

PyTorch MAE / L1 loss function

MAE loss function can be instantiated using `torch.nn.L1Loss`. Similar to MSE loss function, this function calculates different values based on the `reduction` parameter:

```
Loss = nn.L1Loss(reduction='mean')
input = torch.randn(3, 5, requires_grad=True)
target = torch.randn(3, 5)
output = loss(input, target)
```

We can now move on to CE loss, which is used in multi-class classification tasks.

PyTorch CE loss functions

`torch.nn.CrossEntropyLoss` is useful when training a model for a classification problem with multiple classes. As shown in the following code snippet, this class also has a `reduction` parameter for calculating different variations. You can further change the behavior of the loss using `weight` and `ignore_index` parameters, which weight each class and ignore specific indices, respectively:

```
loss = nn.CrossEntropyLoss(reduction="mean")
input = torch.randn(3, 5, requires_grad=True)
target = torch.empty(3, dtype=torch.long).random_(5)
output = loss(input, target)
```

In a similar fashion, we can define BCE loss.

PyTorch BCE loss functions

Similar to CE loss, PyTorch defines the BCE loss as `torch.nn.BCELoss` with the same set of parameters. However, exploiting the close relationship between `torch.nn.BCELoss` and the sigmoid operation, PyTorch provides `torch.nn.BCEWithLogitsLoss`, which achieves higher numerical stability by combining the `softmax` operation and the BCE loss calculation in a single class. The usage is shown in the following code snippet:

```
loss = torch.nn.BCEWithLogitsLoss(reduction="mean")
input = torch.randn(3, requires_grad=True)
target = torch.empty(3).random_(2)
output = loss(input, target)
```

Finally, let's have a look at construction of a custom loss in PyTorch.

PyTorch custom loss functions

Defining a custom loss function is straightforward. Any function defined with PyTorch operations can be used as a loss function.

The following is a sample implementation of `torch.nn.MSELoss` using the mean operator:

```
def custom_mse_loss(output, target):
    loss = torch.mean((output - target)**2)
    return loss
input = torch.randn(3, 5, requires_grad=True)
target = torch.randn(3, 5)
output = custom_mse_loss(input, target)
```

Now, we will move to the overview of optimizers in PyTorch.

PyTorch optimizers

As described in the *PyTorch model training* section, the `configure_optimizers` function of `LightningModule` specifies the optimizer for the training. In PyTorch, predefined optimizers can be found from the `torch.optim` module. The optimizer instantiation requires model parameters, which can be obtained by calling the `parameters` function on the model, as shown in the following sections.

PyTorch SGD optimizer

The following code snippet instantiates an SGD optimizer with an LR of `0.1` and demonstrates how a single step of a model parameter update can be achieved.

`torch.optim.SGD` has built-in support for momentum and acceleration, which further improves training performance. It can be configured using `momentum` and `nesterov` parameters:

```
optimizer = torch.optim.SGD(model.parameters(), lr=0.1
momentum=0.9, nesterov=True)
```

PyTorch Adam optimizer

Similarly, an Adam optimizer can be instantiated using `torch.optim.Adam`, as shown in the following line of code:

```
optimizer = torch.optim.Adam(model.parameters(), lr=0.1)
```

If you are curious about how optimizers work in PyTorch, we recommend reading over the official documentation: `https://pytorch.org/docs/stable/optim.html`.

Things to remember

a. PyTorch is a popular DL framework that provides GPU-accelerated matrix calculation and automatic differentiation. PyTorch is gaining popularity for its flexibility, ease of use, as well as efficiency in model training.

b. For readability and modularity, PyTorch exploits a class called `Dataset` for data management and another class, `DataLoader`, for accessing samples iteratively.

c. The key benefit of PL comes from `LightningModule`, which simplifies the organization of the complex PyTorch code structure into six sections: computation, a train loop, validation loop, test loop, prediction loop, as well as optimizers and LR scheduler

d. PyTorch and PL share the `torch.nn` module for various layers and loss functions. Predefined optimizers can be found from the `torch.optim` module.

In the following section, we will look at another DL framework, TF. Training set up with TF is remarkably similar to the set up with PyTorch.

Implementing and training a model in TF

While PyTorch is oriented towards research projects, TF puts more emphasis on industry use cases. While the deployment features of PyTorch, Torch Serve, and Torch Mobile are still in the experimental phase, the deployment features of TF, TF Serve, and TF Lite are stable and actively in use. The first version of TF was introduced by the Google Brain team in 2011 and they have been continuously updating TF to make it more flexible, user-friendly, and efficient. The key difference between TF and PyTorch was initially much larger, as the first version of TF used static graphs. However, this situation has changed with version 2, as it introduces eager execution, mimicking dynamic graphs known from PyTorch. TF version 2 is often used with **Keras**, an interface for ANN (https://keras.io). Keras allows users to quickly develop DL models and run experiments. In the following sections, we will walk you through the key components of TF.

TF data loading logic

Data can be loaded for TF models in various ways. One of the key data manipulation modules that you should be aware of is `tf.data`, which helps you to build efficient input pipelines. `tf.data` provides `tf.data.Dataset` and `tf.data.TFRecordDataset` classes that are designed for loading datasets of different data formats. In addition, there are `tensorflow_datasets` (tfds) modules (https://www.tensorflow.org/datasets/api_docs/python/tfds) and `tensorflow_addons` modules (https://www.tensorflow.org/addons) that further simplify the data loading process in many cases. It is also worth mentioning the TF I/O package (https://www.tensorflow.org/io/overview), which expands the capabilities of the standard TF file system interaction.

Regardless of the package that you are going to use, you should consider creating a `DataLoader` class. In this class, you will clearly define how the target data will be loaded and how it will be preprocessed before the training. The following code snippet is a sample implementation with loading logic:

```
import tensorflow_datasets as tfds
class DataLoader:
    """ DataLoader class"""
    @staticmethod
    def load_data(config):
        return tfds.load(config.data_url)
```

In the preceding example, we use `tfds` to load data from the external URL (`config.data_url`). More information about `tfds.load` can be found online: https://www.tensorflow.org/datasets/api_docs/python/tfds/load.

Data is available in various formats. Therefore, it is important that it is preprocessed into the format that TF models can consume using the functionalities provided by the `tf.data` module. So, let's have a look at how to use this package for reading data of common formats:

- First, data in `tfrecord`, a format designed for storing a sequence of binary data, can be read as follows:

```
import tensorflow as tf
dataset = tf.data.TFRecordDataset(list_of_files)
```

- We can create a dataset object from a NumPy array using the `tf.data.Dataset.from_tensor_slices` function as follows:

```
dataset = tf.data.Dataset.from_tensor_slices(numpy_array)
```

- Pandas DataFrames can also be loaded as a dataset using the same `tf.data.Dataset.from_tensor_slices` function:

```
dataset = tf.data.Dataset.from_tensor_slices((df_
features.values, df_target.values))
```

- Another option is to use a Python generator. Here is a simple example that highlights how to use a generator to feed a paired image and label:

```
def data_generator(images, labels):
    def fetch_examples():
        i = 0
        while True:
            example = (images[i], labels[i])
            i += 1
            i %= len(labels)
            yield example
        return fetch_examples
training_dataset = tf.data.Dataset.from_generator(
    data_generator(images, labels),
    output_types=(tf.float32, tf.int32),
    output_shapes=(tf.TensorShape(features_shape),
tf.TensorShape(labels_shape)))
```

As shown in the last code snippet, `tf.data.Dataset` provides us with built-in data loading functionalities such as batching, repeating, and shuffling. These options are self-explanatory: batching creates mini-batches of a specific size, repeating allows us to iterate over dataset multiple times, and shuffling mixes up the data entries for every epoch.

Before we wrap up this section, we would like to mention that models implemented with Keras can directly consume NumPy arrays and Pandas DataFrames.

TF model definition

Similar to how PyTorch and PL handles model definition, TF provides various ways of defining network architecture. First, we will look at `Keras.Sequential`, which chains a set of layers to construct a network. This class handles the linkage for you so that you don't need to define the linkage between the layers explicitly:

```
import tensorflow as tf
from tensorflow import keras
from tensorflow.keras import layers
input_shape = 50
model = keras.Sequential(
    [
        keras.Input(shape=input_shape),
        layers.Dense(128, activation="relu", name="layer1"),
        layers.Dense(64, activation="relu", name="layer2"),
        layers.Dense(1, activation="sigmoid", name="layer3"),
    ])
```

In the preceding example, we are creating a model that consists of an input layer, two dense layers, and an output layer that generates a single neuron as an output. This is a simple model that can be used for binary classification.

If the model definition is more complex and cannot be constructed in a sequential manner, another option is to use the `keras.Model` class, as shown in the following code snippet:

```
num_classes = 5
input_1 = layers.Input(50)
input_2 = layers.Input(10)
x_1 = layers.Dense(128, activation="relu", name="layer1x")
(input_1)
x_1 = layers.Dense(64, activation="relu", name="layer1_2x")
(x_1)
```

```
x_2 = layers.Dense(128, activation="relu", name="layer2x")
(input_2)
x_2 = layers.Dense(64, activation="relu", name="layer2_1x")
(x_2)
x = layers.concatenate([x_1, x_2], name="concatenate")
out = layers.Dense(num_classes, activation="softmax",
name="output")(x)
model = keras.Model((input_1,input_2), out)
```

In this example, we have two inputs with a distinct set of computations. The two paths are merged in the last concatenation layer, which transports the concatenated tensor into the final dense layer with five neurons. Given that the last layer uses softmax activation, this model can be used for multi-class classification.

The third option, as follows, is to create a class that inherits keras.Model. This option gives you the most flexibility, as it allows you to customize every part of the model and the training process:

```
class SimpleANN(keras.Model):
    def __init__(self):
        super().__init__()
        self.dense_1 = layers.Dense(128, activation="relu",
name="layer1")
        self.dense_2 = layers.Dense(64, activation="relu",
name="layer2")
        self.out = layers.Dense(1, activation="sigmoid",
name="output")
    def call(self, inputs):
        x = self.dense_1(inputs)
        x = self.dense_3(x)
        return self.out(x)
model = SimpleANN()
```

SimpleANN, from the preceding code, inherits Keras.Model. Within the __init__ function, we need to define the network architecture using a tf.keras.layers module or basic TF operations. The forward propagation logic is defined inside a call method, just as PyTorch has the forward method.

When the model is defined as a distinct class, you can link additional functionalities to the class. In the following example, the `build_graph` method is added to return a `keras.Model` instance, so you can, for example, use the `summary` function to visualize the network architecture as a simpler representation:

```
class SimpleANN(keras.Model):
    def __init__(self):
    ...
    def call(self, inputs):
    ...
    def build_graph(self, raw_shape):
        x = tf.keras.layers.Input(shape=raw_shape)
        return keras.Model(inputs=[x],
outputs=self.call(x))
```

Now, let's look at how TF provides a set of layer implementations through Keras.

TF DL layers

As mentioned in the previous section, the `tf.keras.layers` module provides a set of layer implementations that you can use for building a TF model. In this section, we will cover the same set of layers that we described in the *Implementing and training a model in PyTorch* section. The complete list of layers available in this module can be found at `https://www.tensorflow.org/api_docs/python/tf/keras/layers`.

TF dense (linear) layers

The first one is `tf.keras.layers.Dense`, which performs a linear transformation:

```
tf.keras.layers.Dense(units, activation=None, use_bias=True,
kernel_initializer='glorot_uniform', bias_initializer='zeros',
kernel_regularizer=None, bias_regularizer=None, activity_
regularizer=None, kernel_constraint=None, bias_constraint=None,
**kwargs)
```

The `units` parameter defines the number of neurons in the dense layer (the dimensionality of the output). If the `activation` parameter is not defined, the output of the layer will be returned as is. As presented in the following code, we can apply an `Activation` operation outside of the layer definition as well:

```
X = layers.Dense(128, name="layer2")(input)
x = tf.keras.layers.Activation('relu')(x)
```

In some cases, you will need to build a custom layer. The following example demonstrates how to create a dense layer using basic TF operations by inheriting the `tensorflow.keras.layers.Layer` class:

```python
import tensorflow as tf
from tensorflow.keras.layers import Layer

class CustomDenseLayer(Layer):
    def __init__(self, units=32):
        super(SimpleDense, self).__init__()
        self.units = units
    def build(self, input_shape):
        w_init = tf.random_normal_initializer()
        self.w = tf.Variable(name="kernel",
initial_value=w_init(shape=(input_shape[-1], self.units),

        dtype='float32'),trainable=True)
        b_init = tf.zeros_initializer()
        self.b = tf.Variable(name="bias",initial_value=b_
init(shape=(self.units,), dtype='float32'),trainable=True)
    def call(self, inputs):
        return tf.matmul(inputs, self.w) + self.b
```

Within the `__init__` function of the `CustomDenseLayer` class, we define the dimensionality of the output (`units`). Then, the state of the layer is instantiated within the `build` method; we create and initialize the weights and biases for the layer. The last method, `call`, defines the computation itself. For a dense layer, it consists of multiplying the inputs with the weights and adding biases.

TF pooling layers

`tf.keras.layers` provides different kinds of pooling layers: average, max, global average, and global max pooling layers for one-dimensional temporal data, two-dimensional, or three-dimensional spatial data. In this section, we will show you two-dimensional max pooling and average pooling layers:

```python
tf.keras.layers.MaxPool2D(
    pool_size=(2, 2), strides=None, padding='valid', data_
format=None,
    kwargs)
```

```
tf.keras.layers.AveragePooling2D(
    pool_size=(2, 2), strides=None, padding='valid', data_
format=None,
    kwargs)
```

The two layers both take in `pool_size`, which defines the size of the window. The `strides` parameter is used to define how the windows move throughout the pooling operation.

TF normalization layers

In the following example, we demonstrate a layer for batch normalization, `tf.keras.layers.BatchNormalization`:

```
tf.keras.layers.BatchNormalization(
    axis=-1, momentum=0.99, epsilon=0.001, center=True,
scale=True,
    beta_initializer='zeros', gamma_initializer='ones',
    moving_mean_initializer='zeros',
    moving_variance_initializer='ones', beta_regularizer=None,
    gamma_regularizer=None, beta_constraint=None, gamma_
constraint=None, **kwargs)
```

The output of this layer will have mean close to 0 and standard deviation close to 1. Details about each parameter can be found at `https://www.tensorflow.org/api_docs/python/tf/keras/layers/BatchNormalization`.

TF dropout layers

The `Tf.keras.layers.Dropout` layer applies dropout, a regularization method that sets randomly selected values to zero:

```
tf.keras.layers.Dropout(rate, noise_shape=None, seed=None,
**kwargs)
```

In the preceding layer instantiation, the `rate` argument, a float value between 0 and 1, determines the fraction of the input units that will be dropped.

TF convolution layers

`tf.keras.layers` provides various implementations of convolutional layers, `tf.keras.layers.Conv1D`, `tf.keras.layers.Conv2D`, `tf.keras.layers.Conv3D`, and the corresponding transposed convolutional layers (deconvolution layers), `tf.keras.layers.Conv1DTranspose`, `tf.keras.layers.Conv2DTranspose`, and `tf.keras.layers.Conv3DTranspose`.

The following code snippet describes the instantiation of a two-dimensional convolution layer:

```
tf.keras.layers.Conv2D(
    filters, kernel_size, strides=(1, 1), padding='valid',
    data_format=None, dilation_rate=(1, 1), groups=1,
    activation=None, use_bias=True,
    kernel_initializer='glorot_uniform',
    bias_initializer='zeros', kernel_regularizer=None,
    bias_regularizer=None, activity_regularizer=None,
    kernel_constraint=None, bias_constraint=None, **kwargs)
```

The main parameters in the preceding layer definition are `filters` and `kernel_size`. The `filters` parameter defines the dimensionality of the output and the `kernel_size` parameter defines the size of the two-dimensional convolution window. For the other parameters, please look at https://www.tensorflow.org/api_docs/python/tf/keras/layers/Conv2D.

TF recurrent layers

The following list of recurrent layers is implemented in Keras: the LSTM layer, GRU layer, `SimpleRNN` layer, `TimeDistributed` layer, `Bidirectional` layer, `ConvLSTM2D` layer, and `Base RNN` layer.

In the following code snippet, we demonstrate how to instantiate the `Bidirectional` and LSTM layers:

```
model = Sequential()
model.add(Bidirectional(LSTM(10, return_sequences=True), input_
shape=(5, 10)))
model.add(Bidirectional(LSTM(10)))
model.add(Dense(5))
model.add(Activation('softmax'))
```

In the preceding example, the LSTM layer is modified by a `Bidirectional` wrapper to provide both an initial sequence and a reversed sequence to two copies of the hidden layers. The outputs from the two layers get merged for the final output. By default, the outputs are concatenated but the `merge_mode` parameter allows us to select a different merging option. The dimensionality of the output space is defined by the first parameter. To access the hidden state for each input at every time step, you can enable `return_sequences`. For more details, please look at https://www.tensorflow.org/api_docs/python/tf/keras/layers/LSTM.

TF model training

For Keras models, model training can be achieved by simply calling a `fit` function on the model after calling a `compile` function with an optimizer and a loss function. The `fit` function trains the model using the provided dataset for the given number of epochs.

The following code snippet describes the parameters of the `fit` function:

```
model.fit(
    x=None, y=None, batch_size=None, epochs=1,
    verbose='auto', callbacks=None, validation_split=0.0,
    validation_data=None, shuffle=True,
    class_weight=None, sample_weight=None,
    initial_epoch=0, steps_per_epoch=None,
    validation_steps=None, validation_batch_size=None,
    validation_freq=1, max_queue_size=10, workers=1,
    use_multiprocessing=False)
```

x and y represent the input tensor and the labels. They can be provided in various formats: NumPy arrays, TF tensors, TF datasets, generators, or `tf.keras.utils.experimental.DatasetCreator`. In addition to `fit`, Keras models also have a `train_on_batch` function that only executes a gradient update on a single batch of data.

While TF version 1 requires computation graph compilation for the training loop, TF version 2 allows us to define the training logic without any compilation, as in the case of PyTorch. A typical training loop will look as follows:

```
Optimizer = tf.keras.optimizers.Adam()
loss_fn = tf.keras.losses.CategoricalCrossentropy()
train_acc_metric = tf.keras.metrics.CategoricalAccuracy()
for epoch in range(epochs):
    for step, (x_batch_train, y_batch_train) in enumerate(train_
dataset):
        with tf.GradientTape() as tape:
            logits = model(x_batch_train, training=True)
            loss_value = loss_fn(y_batch_train, logits)
        grads = tape.gradient(loss_value, model.trainable_
weights)
        optimizer.apply_gradients(zip(grads, model.trainable_
weights))
        train_acc_metric.update_state(y, logits)
```

In the preceding code snippet, the outer loop iterates over epochs and the inner loop iterates over the train set. The forward propagation and loss calculation is within the scope of GradientTape, which records operations for automatic differentiation for each batch. Outside of the scope, the optimizer uses the computed gradients to update the weights. In the preceding example, TF functions execute operations immediately, instead of adding the operation to the computation graph, as in eager execution. We would like to mention that you will need to use the @tf.function decorator if you are using TF version 1, where explicit construction of the computation graph is necessary.

Next, we will have a look at loss functions in TF.

TF loss functions

In TF, the loss function needs to be specified when a model is compiled. While you can build a custom loss function from scratch, you can use predefined loss functions provided by Keras through the tf.keras.losses module (https://www.tensorflow.org/api_docs/python/tf/keras/losses). The following example demonstrates how you can use a loss function from Keras to compile a model:

```
model.compile(loss=tf.keras.losses.
BinaryFocalCrossentropy(gamma=2.0, from_logits=True), ...)
```

Additionally, you can pass a string alias to a loss parameter, as shown in the following code snippet:

```
model.compile(loss='sparse_categorical_crossentropy', ...)
```

In this section, we will explain how the loss functions described in the *PyTorch loss functions* section can be instantiated in TF.

TF MSE / L2 loss functions

The MSE / L2 loss function can be defined as follows (https://www.tensorflow.org/api_docs/python/tf/keras/losses/MeanSquaredError):

```
mse = tf.keras.losses.MeanSquaredError()
```

This is the most frequently used loss function for regression – it calculates the mean value of the squared differences between labels and predictions. The default settings will calculate the MSE. However, similar to PyTorch implementation, we can provide a reduction parameter to change that behavior. For example, if you would like to apply a sum operation instead of a mean operation, you can add reduction=tf.keras.losses.Reduction.SUM in the loss function. Given that torch.nn.MSELoss in PyTorch returns the squared difference as is, you can obtain the same loss in TF by passing in reduction=tf.keras.losses.Reduction.NONE to the constructor.

Next, we will look at MAE loss.

TF MAE / L1 loss functions

`tf.keras.losses.MeanAbsoluteError` is the function for MAE loss in Keras (https://www.tensorflow.org/api_docs/python/tf/keras/losses/MeanAbsoluteError):

```
mae = tf.keras.losses.MeanAbsoluteError()
```

As the name suggests, this loss computes the mean of absolute differences between the true and predicted values. It also has a `reduction` parameter that can be used in the same way as described for `tf.keras.losses.MeanSquaredError`.

Now, let's have a look at losses for classification, CE loss.

TF CE loss functions

CE loss calculates the difference between two probability distributions. Keras provides the `tf.keras.losses.CategoricalCrossentropy` class, which is designed for classifying multiple classes (https://www.tensorflow.org/api_docs/python/tf/keras/losses/CategoricalCrossentropy). The following code snippet shows the instantiation:

```
cce = tf.keras.losses.CategoricalCrossentropy()
```

In the case of Keras, labels need to be formatted as one hot vectors. For example, when the target class is the first one out of five classes, it'd be `[1, 0, 0, 0, 0]`.

A CE loss designed for binary classification, BCE loss, also exists.

TF BCE loss functions

In the case of a binary classification, the labels are either 0 or 1. The loss function designed specifically for binary classification, BCE loss, can be defined as follows (https://www.tensorflow.org/api_docs/python/tf/keras/losses/BinaryFocalCrossentropy):

```
loss = tf.keras.losses.BinaryFocalCrossentropy(from_
logits=True)
```

The key parameter for this loss is `from_logits`. When this flag is set to `False`, we have to provide probabilities, continuous values between 0 and 1. When it is set to `True`, we need to provide logits, values between `-infinity` and `+infinity`.

Lastly, let's look at how we can define a custom loss in TF.

TF custom loss functions

To build a custom loss function, we need to create a function that takes predictions and labels as parameters and performs desirable calculations. While TF syntax only expects these two arguments, we can also add some additional arguments by wrapping the function into another function that returns the loss. The following example demonstrates how to create Huber Loss as a custom loss function:

```
def custom_huber_loss(threshold=1.0):
    def huber_fn(y_true, y_pred):
        error = y_true - y_pred
        is_small_error = tf.abs(error) < threshold
        squared_loss = tf.square(error) / 2
        linear_loss = threshold * tf.abs(error) - threshold**2 /
2
        return tf.where(is_small_error, squared_loss, linear_
loss)
    return huber_fn


model.compile(loss=custom_huber_loss (2.0), optimizer="adam"
```

Another option is to create a class that inherits the `tf.keras.losses.Loss` class. We need to implement `__init__` and `call` methods in this case, as follows:

```
class CustomLoss(tf.keras.losses.Loss):
    def __init__(self, threshold=1.0):
        super().__init__()
        self.threshold = threshold
    def call(self, y_true, y_pred):
        error = y_true - y_pred
        is_small_error = tf.abs(error) < threshold
        squared_loss = tf.square(error) / 2
        linear_loss = threshold*tf.abs(error) - threshold**2 / 2
        return tf.where(is_small_error, squared_loss, linear_
loss)


model.compile(optimizer="adam", loss=CustomLoss(),
```

In order to use this loss class, you must instantiate it and pass it to the `compile` function through a `loss` parameter, as described at the beginning of this section.

TF optimizers

In this section, we will describe how to set up different optimizers for model training in TF. Similar to loss functions in the preceding section, Keras provides a set of optimizers for TF through `tf.keras.optimizers`. Out of the various optimizers, we will look at the two main optimizers, SGD and Adam optimizers, in the following section.

TF SGD optimizer

Designed with a fixed LR, an SGD optimizer is the most typical optimizer that you can use for many models. The following code snippet describes how to instantiate an SGD optimizer in TF:

```
tf.keras.optimizers.SGD(
    learning_rate=0.01,
    momentum=0.0,
    nesterov=False,
    name='SGD',
    kwargs)
```

Similar to PyTorch implementation, `tf.keras.optimizers.SGD` also supports an augmented SGD optimizer using the `momentum` and `nesterov` parameters.

TF Adam optimizer

As described in the *Model training logic* section, an Adam optimizer is designed with an adaptive LR. In TF, it can be instantiated as the following:

```
tf.keras.optimizers.Adam(
    learning_rate=0.001, beta_1=0.9, beta_2=0.999,
    epsilon=1e-07, amsgrad=False, name='Adam', **kwargs)
```

For both optimizers, while `learning_rate` plays the most important role of defining the initial LR, we recommend that you review the official documentation to familiarize yourself with the other parameters too: `https://www.tensorflow.org/api_docs/python/tf/keras/optimizers`.

TF callbacks

In this section, we would like to briefly describe callbacks. These are the objects that are used at various stages of training to perform specific actions. The most used callbacks are `EarlyStopping`, `ModelCheckpoint`, and `TensorBoard`, which stop the training when a specific condition is met, save the model after each epoch, and visualize the training status, respectively.

Here is an example of the `EarlyStopping` callback that monitors validation loss and stops the training if the monitored loss has stopped decreasing:

```
tf.keras.callbacks.EarlyStopping(
    monitor='val_loss', min_delta=0.1, patience=2,
    verbose=0, mode='min', baseline=None,
    restore_best_weights=False)
```

The `min_delta` parameter defines the minimum change in the monitored quantity for the change to be considered an improvement and the `patience` parameter defines the number of epochs without any improvements after which the training will be stopped.

Building a custom callback can be achieved by inheriting `keras.callbacks.Callback`. Defining logic for a specific event can be achieved by overwriting its methods, which clearly describe which event it binds to:

- `on_train_begin`
- `on_train_end`
- `on_epoch_begin`
- `on_epoch_end`
- `on_test_begin`
- `on_test_end`
- `on_predict_begin`
- `on_predict_end`
- `on_train_batch_begin`
- `on_train_batch_end`
- `on_predict_batch_begin`
- `on_predict_batch_end`
- `on_test_batch_begin`
- or `on_test_batch_end`

For the complete details, we recommend that you take a look at `https://www.tensorflow.org/api_docs/python/tf/keras/callbacks/Callback`.

> **Things to remember**
>
> a. `tf.data` allows you to build efficient data loading logic. Packages such as `tfds`, `tensorflow addons`, or TF I/O are useful for reading data of different formats.
>
> b. TF, with support from Keras, allows users to construct models using three different approaches: sequential, functional, and subclassing.
>
> c. To simplify model development using TF, the `tf.keras.layers` module provides various layer implementations, the `tf.keras.losses` module includes different loss functions, and the `tf.keras.optimizers` module provides a set of standard optimizers.
>
> d. `Callbacks` can be used to perform specific actions at the various stages of training. The commonly used callbacks are `EarlyStopping` and `ModelCheckpoint`.

So far, we have learned how to set up a DL model training using the most popular DL frameworks, PyTorch and TF. In the following section, we will look at how the components that we have described in this section are used in reality.

Decomposing a complex, state-of-the-art model implementation

Even though you have picked up the basics of TF and PyTorch, setting up a model training from scratch can be overwhelming. Luckily, the two frameworks have thorough documentations and tutorials that are easy to follow:

- TF

  - Image classification with convolution layers: `https://www.tensorflow.org/tutorials/images/classification`.

  - Text classification with recurrent layers: `https://www.tensorflow.org/text/tutorials/text_classification_rnn`.

- PyTorch

  - Object detection with convolutional layers: `https://pytorch.org/tutorials/intermediate/torchvision_tutorial.html`.

  - Machine translation with recurrent layers: `https://pytorch.org/tutorials/intermediate/seq2seq_translation_tutorial.html`.

In this section, we would like to look at a model that is much more sophisticated, StyleGAN. Our main goal is to explain how the components described in the previous sections can be put together for a complex DL project. For the complete description of the model architecture and performance, we recommend the publication released by NVIDIA, available at `https://ieeexplore.ieee.org/document/8953766`.

StyleGAN

StyleGAN, as a variation of a **generative adversarial network (GAN)**, aims to generate new images from latent codes (random noise vectors). Its architecture can be broken down into three elements: a mapping network, a generator, and a discriminator. At a high level, the mapping network and generator work together to generate an image from a set of random values. The discriminator plays a critical role of guiding the generator to generate realistic images during training. Let's take a closer look at each component.

The mapping network and generator

While generators are designed to process latent codes directly in a traditional GAN, latent codes are fed to the mapping network first in StyleGAN, as shown in *Figure 3.5*. The output of the mapping network is then fed to each step of the generator, changing the style and details of the generated image. The generator starts at a lower resolution, constructing outlines for the image at a tensor size of 4 x 4 or 8 x 8. The details of the images are filled as the generator handles the bigger tensors. At the last couple of layers, the generator interacts with tensors of sizes 64 x 64 and 1024 x 1024 to construct the high-resolution features:

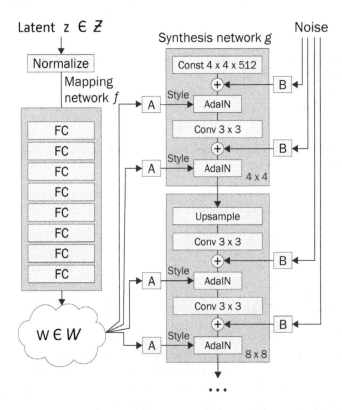

Figure 3.5 – A mapping network (left) and generator (right) of StyleGAN

In the preceding figure, the network that takes in a latent vector, **z**, and generates **w** is the mapping network. The network on the right is the generator, **g**, which takes in a set of noise vectors, as well as **w**. The discriminator is fairly simple compared to the generator. The layers are depicted in *Figure 3.6*:

Discriminator	Activation	Output Shape	Params
Input image	–	3 x 1024 x 1024	–
Conv 1 x 1	LReLU	16 x 1024 x 1024	64
Conv 3 x 3	LReLU	16 x 1024 x 1024	2.3k
Conv 3 x 3	LReLU	32 x 1024 x 1024	4.6k
Downsample	–	32 x 512 x 512	–
Conv 3 x 3	LReLU	32 x 512 x 512	9.2k
Conv 3 x 3	LReLU	64 x 512 x 512	18k
Downsample	–	64 x 256 x 256	–
Conv 3 x 3	LReLU	64 x 256 x 256	37k
Conv 3 x 3	LReLU	128 x 256 x 256	74k
Downsample	–	128 x 128 x 128	–
Conv 3 x 3	LReLU	128 x 128 x 128	148k
Conv 3 x 3	LReLU	256 x 128 x 128	295k
Downsample	–	256 x 64 x 64	–
Conv 3 x 3	LReLU	256 x 64 x 64	590k
Conv 3 x 3	LReLU	512 x 64 x 64	1.2M
Downsample	–	512 x 32 x 32	–
Conv 3 x 3	LReLU	512 x 32 x 32	2.4M
Conv 3 x 3	LReLU	512 x 32 x 32	2.4M
Downsample	–	512 x 16 x 16	–
Conv 3 x 3	LReLU	512 x 16 x 16	2.4M
Conv 3 x 3	LReLU	512 x 16 x 16	2.4M
Downsample	–	512 x 8 x 8	–
Conv 3 x 3	LReLU	512 x 8 x 8	2.4M
Conv 3 x 3	LReLU	512 x 8 x 8	2.4M
Downsample	–	512 x 4 x 4	–
Minibatch stddev	–	513 x 4 x 4	–
Conv 3 x 3	LReLU	512 x 4 x 4	2.4M
Conv 4 x 4	LReLU	512 x 1 x 1	4.2M
Fully-connected	linear	1 x 1 x 1	513
Total trainable parameters			23.1M

Figure 3.6 – A StyleGAN discriminator architecture for the FFHQ dataset at 1024 × 1024 resolution

As depicted in the preceding image, the discriminator consists of multiple blocks of convolution layers and downsampling operations. It takes in an image of size 1024 x 1024 and generates a numeric value between 0 and 1, describing how realistic the image is.

Training StyleGAN

Training StyleGAN requires a lot of computations, so multiple GPUs are necessary to achieve a reasonable training time. The estimations are summarized in *Figure 3.7*:

GPUs	1024x1024	512x512
1	41 days 4 hours	24 days 21 hours
2	21 days 22 hours	13 days 7 hours
4	11 days 8 hours	7 days 0 hours
8	6 days 14 hours	4 days 10 hours

Figure 3.7 – The training time for StyleGAN with an FFHQ dataset on Tesla V100 GPUs

Therefore, if you want to play around with StyleGAN, we recommend following the instructions in the official GitHub repositories, where they provide pre-trained models: `https://github.com/NVlabs/stylegan`.

Implementation in PyTorch

Unfortunately, NVIDIA has not shared the public implementation of StyleGAN in PyTorch. Instead, they have released StyleGAN2, which shares most of the same components. Therefore, we will use the StyleGAN2 implementation for our PyTorch example: `https://github.com/NVlabs/stylegan2-ada-pytorch`.

All the network components are found under `training/network.py`. The three components are named as described in the previous section: `MappingNetwork`, `Generator`, and `Discriminator`.

The mapping network in PyTorch

The implementation of `MappingNetwork` is self-explanatory. The following code snippet includes the core logic for the mapping network:

```
class MappingNetwork(torch.nn.Module):
    def __init__(self, ...):
        ...
        for idx in range(num_layers):
```

```
            in_features = features_list[idx]
            out_features = features_list[idx + 1]
            layer = FullyConnectedLayer(in_features,
   out_features, activation=activation, lr_multiplier=
   lr_multiplier) setattr(self, f'fc{idx}', layer)


        def forward(self, z, ...):
            # Embed, normalize, and concat inputs.
            x = normalize_2nd_moment(z.to(torch.float32))


            # Main layers
            for idx in range(self.num_layers):
                layer = getattr(self, f'fc{idx}')
                x = layer(x)
            return x
```

In this network definition, `MappingNetwork` inherits `torch.nn.Module`. Within the `__init__` function, the necessary `FullyConnectedLayer` instances are initialized. The `forward` method feeds the latent vector, z, to each layer.

The generator in PyTorch

The following code snippet describes how the generator is implemented. It consists of `MappingNetwork` and `SynthesisNetwork`, as depicted in *Figure 3.5*:

```
class Generator(torch.nn.Module):
    def __init__(self, …):
        self.z_dim = z_dim
        self.c_dim = c_dim
        self.w_dim = w_dim
        self.img_resolution = img_resolution
        self.img_channels = img_channels
        self.synthesis = SynthesisNetwork(
            w_dim=w_dim,
            img_resolution=img_resolution,
            img_channels=img_channels,
            synthesis_kwargs)
```

```
        self.num_ws = self.synthesis.num_ws
        self.mapping = MappingNetwork(
            z_dim=z_dim, c_dim=c_dim, w_dim=w_dim,
            num_ws=self.num_ws, **mapping_kwargs)
    def forward(self, z, c, truncation_psi=1, truncation_
cutoff=None, **synthesis_kwargs):
        ws = self.mapping(z, c,
        truncation_psi=truncation_psi,
        truncation_cutoff=truncation_cutoff)
        img = self.synthesis(ws, **synthesis_kwargs)
        return img
```

The generator network, `Generator`, also inherits `torch.nn.Module`. `SynthesisNetwork` and `MappingNetwork` are instantiated within the `__init__` function and get triggered sequentially in the `forward` function. The implementation of `SynthesisNetwork` is summarized in the following code snippet:

```
class SynthesisNetwork(torch.nn.Module):
    def __init__(self, ...):
        for res in self.block_resolutions:
            block = SynthesisBlock(
                in_channels, out_channels, w_dim=w_dim,
                resolution=res, img_channels=img_channels,
                is_last=is_last, use_fp16=use_fp16,
                block_kwargs)
            setattr(self, f'b{res}', block)
        ...
    def forward(self, ws, **block_kwargs):
        ...
        x = img = None
        for res, cur_ws in zip(self.block_resolutions, block_
ws):
            block = getattr(self, f'b{res}')
            x, img = block(x, img, cur_ws, **block_kwargs)
        return img
```

SynthesisNetwork has multiple blocks of SynthesisBlock. SynthesisBlock receives noise vectors and the output of MappingNetwork to generate a tensor that eventually becomes the output image.

The discriminator in PyTorch

The following code snippet summarizes the PyTorch implementation of Discriminator. The network architecture follows the structure depicted in *Figure 3.6*:

```
class Discriminator(torch.nn.Module):
    def __init__(self, ...):
        self.block_resolutions = [2 ** i for i in range(self.
img_resolution_log2, 2, -1)]
        for res in self.block_resolutions:
            block = DiscriminatorBlock(
                in_channels, tmp_channels, out_channels,
                resolution=res,
                first_layer_idx = cur_layer_idx,
                use_fp16=use_fp16, **block_kwargs,
                common_kwargs)
            setattr(self, f'b{res}', block)
    def forward(self, img, c, **block_kwargs):
        x = None
        for res in self.block_resolutions:
            block = getattr(self, f'b{res}')
            x, img = block(x, img, **block_kwargs)
        return x
```

Similar to SynthesisNetwork, Discriminator makes use of the DiscriminatorBlock class to dynamically create a set of convolutional layers of different sizes. They are defined in the __init__ function, and the tensors are fed to each block sequentially in the forward function.

Model training logic in PyTorch

Training logic is defined in the training_loop function in training/train_loop.py. The original implementation contains a lot of details. In the following code snippet, we will look at the main components that align with what we have learned in the *PyTorch model training* section:

```
def training_loop(...):
    ...
training_set_iterator = iter(torch.utils.data.
```

```
DataLoader(dataset=training_set, sampler=training_set_sampler,
batch_size=batch_size//num_gpus, **data_loader_kwargs))
    loss = dnnlib.util.construct_class_by_name(device=device,
**ddp_modules, **loss_kwargs) # subclass of training.loss.Loss
    while True:
        # Fetch training data.
        with torch.autograd.profiler.record_function('data_
fetch'):
            phase_real_img, phase_real_c = next(training_set_
iterator)
        # Execute training phases.
        for phase, phase_gen_z, phase_gen_c in zip(phases, all_
gen_z, all_gen_c):
            # Accumulate gradients over multiple rounds.
            for round_idx, (real_img, real_c, gen_z, gen_c) in
enumerate(zip(phase_real_img, phase_real_c, phase_gen_z, phase_
gen_c)):
                loss.accumulate_gradients(phase=phase.name, real_
img=real_img, real_c=real_c, gen_z=gen_z, gen_c=gen_c,
sync=sync, gain=gain)
            # Update weights.
            phase.module.requires_grad_(False)
            with torch.autograd.profiler.record_function(phase.name
+ '_opt'):
                phase.opt.step()
```

This function receives configurations for various training components and trains both `Generator` and `Discriminator`. The outer loop iterates over training samples, and the inner loop handles gradient calculation and model parameter updates. The training settings are configured by a separate script, `main/train.py`.

This summarizes the structure of PyTorch implementation. Even though the repository looks overwhelming due to the large number of files, we have walked you through how to break the implementation down into the components that we have described in the *Implementing and training a model in PyTorch* section. In the following section, we will look at implementation in TF.

Implementation in TF

Even though the official implementation is in TF (https://github.com/NVlabs/stylegan), we will look at a different implementation presented in *Hands-On Image Generation with TensorFlow: A Practical Guide to Generating Images and Videos Using Deep Learning* by Soon Yau Cheong. This version is based on TF version 2 and aligns better with what we have described in this book. The implementation can be found at https://github.com/PacktPublishing/Hands-On-Image-Generation-with-TensorFlow-2.0/blob/master/Chapter07/ch7_faster_stylegan.ipynb.

Similar to the PyTorch implementation described in the previous section, the original TF implementation consists of G_mapping for the mapping network, G_style for the generator, and D_basic for the discriminator.

The mapping network in TF

Let's look at the mapping network defined at https://github.com/NVlabs/stylegan/blob/1e0d5c781384ef12b50ef20a62fee5d78b38e88f/training/networks_stylegan.py#L384 and its TF version 2 implementation shown below:

```
def Mapping(num_stages, input_shape=512):
    z = Input(shape=(input_shape))
    w = PixelNorm()(z)
    for i in range(8):
        w = DenseBlock(512, lrmul=0.01)(w)
        w = LeakyReLU(0.2)(w)
    w = tf.tile(tf.expand_dims(w, 1), (1,num_stages,1))
    return Model(z, w, name='mapping')
```

The implementation of MappingNetwork is almost self-explanatory. We can see that the mapping network starts with vector w constructed from a latent vector, z, using a PixelNorm custom layer. The custom layer is defined as follows:

```
class PixelNorm(Layer):
    def __init__(self, epsilon=1e-8):
        super(PixelNorm, self).__init__()
        self.epsilon = epsilon
    def call(self, input_tensor):
        return input_tensor / tf.math.sqrt(tf.reduce_mean(input_
tensor**2, axis=-1, keepdims=True) + self.epsilon)
```

As described in the *TF dense (linear) layers* section, `PixelNorm` inherits the `tensorflow.keras.layers.Layer` class and defines the computation within the `call` function.

The remaining components of `Mapping` are a set of dense layers with `LeakyReLU` activations.

Next, we will have a look at the generator network.

The generator in TF

The generator in the original code, `G_style`, is composed of two networks: `G_mapping` and `G_synthesis`. See the following: `https://github.com/NVlabs/stylegan/blob/1e0d5c781384ef12b50ef20a62fee5d78b38e88f/training/networks_stylegan.py#L299`.

The complete implementation from the repository might look extremely complex at first. However, you will soon find out that `G_style` simply calls `G_mapping` and `G_synthesis` sequentially.

The implementation of `SynthesisNetwork` is summarized in the following code snippet: `https://github.com/NVlabs/stylegan/blob/1e0d5c781384ef12b50ef20a62fee5d78b38e88f/training/networks_stylegan.py#L440`.

In TF version 2, the generator is implemented as follows:

```python
def GenBlock(filter_num, res, input_shape, is_base):
    input_tensor = Input(shape=input_shape, name=f'g_{res}')
    noise = Input(shape=(res, res, 1), name=f'noise_{res}')
    w = Input(shape=512)
    x = input_tensor
    if not is_base:
        x = UpSampling2D((2,2))(x)
        x = ConvBlock(filter_num, 3)(x)
    x = AddNoise()([x, noise])
    x = LeakyReLU(0.2)(x)
    x = InstanceNormalization()(x)
    x = AdaIN()([x, w])
    # Adding noise
    x = ConvBlock(filter_num, 3)(x)
    x = AddNoise()([x, noise])
    x = LeakyReLU(0.2)(x)
```

```
    x = InstanceNormalization()(x)
    x = AdaIN()([x, w])
    return Model([input_tensor, w, noise], x, name=f'genblock_
{res}x{res}')
```

This network follows the architecture depicted in *Figure 3.5*; SynthesisNetwork is constructed with a set of AdaIn and ConvBlock custom layers.

Let's move on to the discriminator network.

The discriminator in TF

The D_basic function implements the discriminator depicted in *Figure 3.6*. (https://github.com/NVlabs/stylegan/blob/1e0d5c781384ef12b50ef20a62fee5d78b38e88f/training/networks_stylegan.py#L562). Since the discriminator consists of a set of convolution layer blocks, D_basic has a dedicated function, block, that builds a block based on the input tensor size. The core components of the function look as follows:

```
def block(x, res): # res = 2 … resolution_log2
    with tf.variable_scope('%dx%d' % (2**res, 2**res)):
        x = act(apply_bias(conv2d(x, fmaps=nf(res-1), kernel=3,
gain=gain, use_wscale=use_wscale)))
        x = act(apply_bias(conv2d_downscale2d(blur(x),
fmaps=nf(res-2), kernel=3, gain=gain, use_wscale=use_wscale,
fused_scale=fused_scale)))
    return x
```

In the preceding code, the block function deals with creating each block in the discriminator by combining convolution and downsampling layers. The remaining logic of D_basic is straightforward, as it simply chains a set of convolution layer blocks by passing the output of one block as an input to the next block.

Model training logic in TF

The training logic for TF implementation can be found in the train_step function. Understanding the implementation details should not be challenging as they have followed the description we had in the *TF model training* section.

Overall, we have learned how StyleGAN can be implemented in TF version 2 using the TF building blocks that we described in this chapter.

> **Things to remember**
>
> a. Any DL model training implementation can be broken into three components (data loading logic, model definition, and model training logic), regardless of the complexity of the implementation.

At this stage, you should understand how the StyleGAN repository is structured in each framework. We strongly recommend that you play around with the pre-trained models to generate interesting images. If you master StyleGAN, it should be easy to follow the implementation of StyleGAN2 (`https://arxiv.org/abs/1912.04958`), StyleGAN3 (`https://arxiv.org/abs/2106.12423`), and HyperStyle (`https://arxiv.org/abs/2111.15666`).

Summary

In this chapter, we have explored where the flexibility of DL comes from. DL uses a network of mathematical neurons to learn the hidden patterns within a set of data. Training a network involves the iterative process of updating model parameters based on a train set and selecting the model that performs the best on a validation set, with the goal of producing the best performance on a test set.

Realizing the repeated processes within model training, many engineers and researchers have put together common building blocks into frameworks. We have described two of the most popular frameworks: PyTorch and TF. The two frameworks are structured in a similar way, allowing users to set up the model training using three building blocks: data loading logic, model definition, and model training logic. As the final topic of the chapter, we decomposed StyleGAN, one of the most popular GAN implementations, to understand how the building blocks are used in reality.

As DL requires a large amount of data for successful training, efficient management of the data, model implementations, and various training results are critical to the success of any project. In the following chapter, we will introduce useful tools for DL experiment monitoring.

4

Experiment Tracking, Model Management, and Dataset Versioning

In this chapter, we will introduce a set of useful tools for experiment tracking, model management, and dataset versioning, which allows you to effectively manage **deep learning** (**DL**) projects. The tools we will be discussing in this chapter can help us track many experiments and interpret the results more efficiently, which naturally leads to a reduction in operational costs and boosts the development cycle. By the end of the chapter, you will have hands-on experience with the most popular tools and be able to select the right set of tools for your project.

In this chapter, we're going to cover the following main topics:

- Overview of DL project tracking
- DL project tracking with Weights & Biases
- DL project tracking with MLflow and DVC
- Dataset versioning – beyond Weights & Biases, MLflow, and DVC

Technical requirements

You can download the supplemental material for this chapter from this book's GitHub repository at `https://github.com/PacktPublishing/Production-Ready-Applied-Deep-Learning/tree/main/Chapter_4`.

Overview of DL project tracking

Training DL models is an iterative process that consumes a lot of time and resources. Therefore, keeping track of all experiments and consistently organizing them can prevent us from wasting our time on unnecessary operations such as training similar models repeatedly on the same set of data. In other words, having well-documented records of all model architectures and their hyperparameter sets, as well as the version of data used during experiments, can help us derive the right conclusion from the experiments, which naturally leads to the project being successful.

Components of DL project tracking

The essential components of **DL project tracking** are **experiment tracking**, **model management**, and **dataset versioning**. Let's look at each component in detail.

Experiment tracking

The concept behind experiment tracking is simple: store the description and the motivations of each experiment so that we don't run another set of experiments for the same purpose. Overall, effective experiment tracking will save us operational costs and allows us to derive the right conclusion from a minimal set of experimental results. One of the basic approaches for effective experiment tracking is adding a unique identifier to each experiment. The information we need to track for each experiment includes project dependencies, the definition of the model architecture, parameters used, and evaluation metrics. Experiment tracking also includes visualizing ongoing experiments in real time and being able to compare a set of experiments intuitively. For example, if we can check train and validation losses from every epoch as the model gets trained, we can identify overfitting quicker, saving some resources. Also, by comparing results and a set of changes made between two experiments, we can understand how the changes affect the model performance.

Model management

Model management goes beyond experiment tracking as it covers the full life cycle of a model: dataset information, artifacts (any data generated from training a model), the implementation of the model, evaluation metrics, and pipeline information (such as development, testing, staging, and production). Model management allows us to quickly pick up the model of interest and efficiently set up the environment in which the model can be used.

Dataset versioning

The last component of DL project tracking is dataset versioning. In many projects, datasets change over time. Changes can come from data schemas (blueprints of how the data is organized), file locations, or even from filters applied to the dataset manipulating the meaning of the underlying data. Many datasets found in the industry are structured in a complex way and often stored in multiple locations in various data formats. Therefore, changes can be more dramatic and harder to track than you anticipated. As a result, keeping a record of the changes is critical in reproducing consistent results throughout the project.

Dataset tracking can be summarized as follows: a set of data stored as an artifact should become a new version of the artifact whenever the underlying data is modified. Having said that, every artifact should have metadata that consists of important information about the dataset: when it is created, who created it, and how it is different from the previous version.

For example, a dataset with dataset versioning should be formulated as follows. The dataset should have a timestamp in its name:

```
dataset_<timestamp>
> metadata.json
> img1.png
> img2.png
> img3.png
```

As mentioned previously, the metadata should contain key information about the dataset:

```
{
    "created_by": "Adam"
    "created_on": "2022-01-01"
    "labelled_by": "Bob"
    "number_of_samples": 3
}
```

Please note that the set of information that's tracked by metadata may be different for each project.

Tools for DL project tracking

DL tracking can be achieved in various ways, starting from simple notes in a text file, through spreadsheets, keeping the information in GitHub or dedicated web pages, to self-built platforms and external tools. Model and data artifacts can be stored as is, or more sophisticated methods can be applied to avoid redundancy and increase efficiency.

The field of DL project tracking is growing fast and is introducing new tools continuously. As a result, selecting the right tool for the underlying project is not an easy task. We must consider both business and technical constraints. While the pricing model is a basic one, the other constraints can possibly be introduced by the existing development settings; integrating the existing tools should be easy, and the infrastructure must be easy to maintain. It is also important to consider the engineering competence of the MLOps team. Having said that, the following list would be a good starting point when you're selecting a tool for your project.

- TensorBoard (https://www.tensorflow.org/tensorboard):
  - An open source visualization tool developed by the TensorFlow team

- A standard tool for tracking and visualizing the experimental results

- Weights & Biases (`https://wandb.ai`):

  - A cloud-based service with an effective and interactive dashboard for visualizing and organizing the experimental results

  - The server can be run locally or hosted in a private cloud

  - It provides an automated hyperparameter-tuning feature called Sweeps

  - Free for personal projects. Pricing is based on the tracking hours and storage space

- Neptune (`https://neptune.ai`):

  - An online tool for monitoring and storing the artifacts from machine learning (ML) experiments

  - It can easily be integrated with the other ML tools

  - It's known for its powerful dashboard which summarizes the experiments in real time

- MLflow (`https://mlflow.org`):

  - An open source platform that offers end-to-end ML life cycle management

  - It supports both Python and R-based systems. It is often used in combination with **Data Version Control (DVC)**

- SageMaker Studio (`https://aws.amazon.com/sagemaker/studio/`):

  - A web-based visual interface for managing ML experiments set up with SageMaker

  - The tool allows users to efficiently build, train, and deploy models by providing simple integrations to the other useful features of AWS

- Kubeflow (`https://www.kubeflow.org`):

  - An open source platform designed by Google for end-to-end ML orchestration and management

  - It is also designed for deploying ML systems to various development and production environments efficiently

- Valohai (`https://valohai.com`):

  - A DL management platform designed for automatic machine orchestration, version control, and data pipeline management

  - It is not free software as it's designed for an enterprise

  - It is gaining popularity for being technology agnostic and having a responsive support team

Out of the various tools, we will cover the two most commonly used settings: Weights & Biases and MLflow combined with DVC.

Things to remember

a. The essential components of DL tracking are experiment tracking, model management, and dataset versioning. Recent DL tracking tools often have user-friendly dashboards that summarize the experimental results.

b. The field is growing and there are many tools with different advantages. Selecting the right tool involves understanding both business and technical constraints.

First, let's look at DL project tracking with **Weights & Biases (W&B)**.

DL project tracking with Weights & Biases

W&B is an experiment management platform that provides versioning for models and data.

W&B provides an interactive dashboard that can be embedded in Jupyter notebooks or used as a standalone web page. The simple Python API opens up the possibility for simple integration as well. Furthermore, its features focus on simplifying DL experiment management: logging and monitoring model and data versions, hyperparameter values, evaluation metrics, artifacts, and other related information.

Another interesting feature of W&B is its built-in hyperparameter search feature called **Sweeps** (`https://docs.wandb.ai/guides/sweeps`). Sweeps can easily be set up using the Python API, and the results and models can be compared interactively on the W&B web page.

Finally, W&B automatically creates reports for you that summarize and organize a set of experiments intuitively (`https://docs.wandb.ai/guides/reports`).

Overall, the key functionalities of W&B can be summarized as follows:

- **Experiment tracking and management**
- **Artifact management**
- **Model evaluation**
- **Model optimization**
- **Collaborative analysis**

W&B is a subscription-based service, but personal accounts are free of charge.

Setting up W&B

W&B has a Python API that provides simple integration methods for many DL frameworks, including TensorFlow and PyTorch. The logged information, such as projects, teams, and the list of runs, is managed and visible online or on a self-hosted server.

The first step of setting up W&B is to install the Python API and log into the W&B server. You must create an account beforehand through https://wandb.ai:

```
pip install wandb
wandb login
```

Within your Python code, you can register a single experiment that will be called run-1 through the following line of code:

```
import wandb
run_1 = wandb.init(project="example-DL-Book", name="run-1")
```

More precisely, the wandb.init function creates a new wandb.Run instance named run_1 within a project called example-DL-Book. If a name is not provided, W&B will generate a random two-word name for you. If the project name is empty, W&B will put your run into the Uncategorized project. All the parameters of wandb.init are listed at https://docs.wandb.ai/ref/python/init, but we would like to introduce the ones that you will mostly interact with:

- id sets a unique ID for your run

- resume allows you to resume an experiment without creating a new run

- job_type allows you to assign your run to a specific type such as training, testing, validation, exploration, or any other name that can be used for grouping the runs

- tags gives you additional flexibility for organizing your runs

When the wandb.init function is triggered, information about the run will start appearing on the W&B dashboard. You can monitor the dashboard on the W&B web page or directly in the Jupyter notebook environment, as shown in the following screenshot:

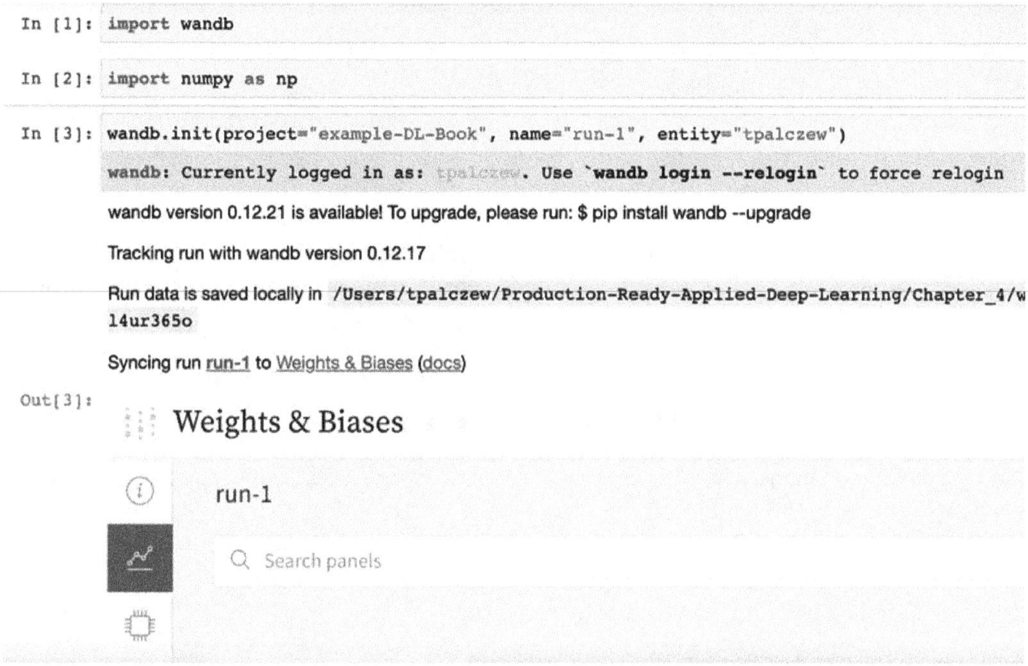

```
In [1]:  import wandb
```

```
In [2]:  import numpy as np
```

```
In [3]:  wandb.init(project="example-DL-Book", name="run-1", entity="tpalczew")
```

wandb: Currently logged in as: tpalczew. Use `wandb login --relogin` to force relogin

wandb version 0.12.21 is available! To upgrade, please run: $ pip install wandb --upgrade

Tracking run with wandb version 0.12.17

Run data is saved locally in /Users/tpalczew/Production-Ready-Applied-Deep-Learning/Chapter_4/w 14ur365o

Syncing run run-1 to Weights & Biases (docs)

Out[3]:

⸬⸬ **Weights & Biases**

ⓘ   run-1

     Q Search panels

Figure 4.1 – The W&B dashboard inside a Jupyter notebook environment

When the run is created, you can start logging information; the wandb.log function allows you to log any data you want. For example, you can log loss during training by adding wandb. log({"custom_loss": custom_loss}) to the training loop. Similarly, you can log validation loss and any other details that you want to keep track of.

Interestingly, W&B made this process even simpler by providing built-in logging functionalities for DL models. At the time of writing, you can find integrations for most frameworks, including Keras, PyTorch, PyTorch Lightning, TensorFlow, fast.ai, scikit-learn, SageMaker, Kubeflow, Docker, Databricks, and Ray Tune (for details, see https://docs.wandb.ai/guides/integrations).

wandb.config is an excellent place to track model hyperparameters. For any artifacts from experiments, you can use the wandb.log_artifact method (for more details, see https:// docs.wandb.ai/guides/artifacts). When logging an artifact, you need to define a file path and then assign the name and type of your artifact, as shown in the following code snippet:

```
wandb.log_artifact(file_path, name='new_artifact', type='my_
dataset')
```

Then, you can reuse the artifact that's been stored, as follows:

```
run = wandb.init(project="example-DL-Book")
artifact = run.use_artifact('example-DL-Book/new_artifact:v0',
type='my_dataset')
artifact_dir = artifact.download()
```

So far, you have learned how to set up wandb for your project and log metrics and artifacts of your choice individually throughout training. Interestingly, wandb provides automatic logging for many DL frameworks. In this chapter, we will take a closer look at W&B integration for Keras and **PyTorch Lighting (PL)**.

Integrating W&B into a Keras project

In the case of Keras, integration can be achieved through the WandbCallback class. The complete version can be found in this book's GitHub repository:

```
import wandb
from wandb.keras import WandbCallback
from tensorflow import keras
from tensorflow.keras import layers
wandb.init(project="example-DL-Book", name="run-1")
wandb.config = {
    "learning_rate": 0.001,
    "epochs": 50,
    "batch_size": 128
}
model = keras.Sequential()
logging_callback = WandbCallback(log_evaluation=True)
model.fit(
    x=x_train, y=y_train,
    epochs=wandb.config['epochs'],
    batch_size=wandb.config['batch_size'],
    verbose='auto',
    validation_data=(x_valid, y_valid),
    callbacks=[logging_callback])
```

As described in the previous section, key information about the models gets logged and becomes available on the W&B dashboard. You can monitor losses, evaluation metrics, and hyperparameters. *Figure 4.2* shows the sample plots that are generated automatically by W&B through the preceding code:

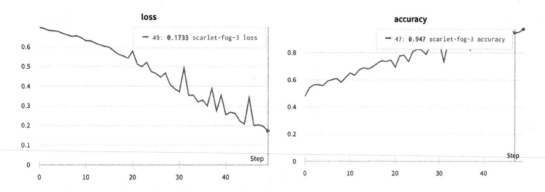

Figure 4.2 – Sample plots generated by W&B from logged metrics

Integrating W&B into a PL project is similar to integrating W&B into a Keras project.

Integrating W&B into a PyTorch Lightning project

For a project based on PL, W&B provides a custom logger and hides most of the boilerplate code. All you need to do is instantiate the `WandbLogger` class and pass it to the `Trainer` instance through `logger` parameter:

```
import pytorch_lightning as pl
from pytorch_lightning.loggers import WandbLogger
wandb_logger = WandbLogger(project="example-DL-Book")
trainer = Trainer(logger=wandb_logger)
class LitModule(LightningModule):
    def __init__(self, *args, **kwarg):
        self.save_hyperparameters()
    def training_step(self, batch, batch_idx):
        self.log("train/loss", loss)
```

A detailed explanation of the integration can be found at `https://pytorch-lightning.readthedocs.io/en/stable/extensions/generated/pytorch_lightning.loggers.WandbLogger.html`.

> **Things to remember**
>
> a. W&B is an experiment management platform that helps in tracking different versions of models and data. It also supports storing configurations, hyperparameters, data, and model artifacts while providing experiment tracking in real time.
>
> b. W&B is easy to set up. It provides a built-in integration feature for many DL frameworks, including TensorFlow and PyTorch.
>
> c. W&B can be used to perform hyperparameter tuning/model optimization.

While W&B has been dominating the field of DL project tracking, the combination of MLflow and DVC is another popular setup for a DL project.

DL project tracking with MLflow and DVC

MLflow is a popular framework that supports tracking technical dependencies, model parameters, metrics, and artifacts. The key components of MLflow are as follows:

- **Tracking**: It keeps a track of result changes every time the model runs
- **Projects**: It packages model code in a reproducible way
- **Models**: It organizes model artifacts for future convenient deployments
- **Model Registry**: It manages a full life cycle of an MLflow model
- **Plugins**: It can be easily integrated with other DL frameworks as it provides flexible plugins

As you may have already noticed, there are some similarities between W&B and MLflow. However, in the case of MLflow, every experiment is linked with a set of Git commits. Git does not prevent us from saving datasets, but it shows many limitations when the datasets are large, even with an extension built for large files (Git LFS). Thus, MLflow is commonly combined with DVC, an open source version control system that solves Git limitations.

Setting up MLflow

MLflow can be installed using `pip`:

```
pip install mlflow
```

Similar to W&B, MLflow also provides a Python API that allows you to track hyperparameters (`log_param`), evaluation metrics (`log_metric`), and artifacts (`log_artifacts`):

```
import os
import mlflow
from mlflow import log_metric, log_param, log_artifacts
log_param("epochs", 30)
log_metric("custom", 0.6)
log_metric("custom", 0.75) # metrics can be updated
if not os.path.exists("artifact_dir"):
    os.makedirs("artifact_dir")
with open("artifact_dir/test.txt", "w") as f:
    f.write("simple example")
log_artifacts("artifact_dir")
```

The experiment definition can be initialized and tagged with the following code:

```
exp_id = mlflow.create_experiment("DLBookModel_1")
exp = mlflow.get_experiment(exp_id)
with mlflow.start_run(experiment_id=exp.experiment_id, run_
name='run_1') as run:
    # logging starts here
    mlflow.set_tag('model_name', 'model1_dev')
```

MLflow has provided a set of tutorials that introduce its APIs: https://www.mlflow.org/docs/latest/tutorials-and-examples/tutorial.html.

Now that you are familiar with the basic usage of MLflow, we will describe how it can be integrated for Keras and PL projects.

Integrating MLflow into a Keras project

First, let's take a look at Keras integration. Logging the details of a Keras model using MLflow can be achieved through the log_model function:

```
history = keras_model.fit(...)
mlflow.keras.log_model(keras_model, model_dir)
```

The mlflow.keras and mlflow.tensorflow modules provide a set of APIs for logging various information about Keras and TensorFlow models, respectively. For additional details, please look at https://www.mlflow.org/docs/latest/python_api/index.html.

Integrating MLflow into a PyTorch Lightning project

Similar to how W&B supports PL projects, MLflow also provides an MLFlowLogger class. This can be passed to a Trainer instance for logging the model details in MLflow:

```
import pytorch_lightning as pl
from pytorch_lightning import Trainer
from pytorch_lightning.loggers import MLFlowLogger
mlf_logger = MLFlowLogger(experiment_name="example-DL-Book ",
tracking_uri="file:./ml-runs")
trainer = Trainer(logger=mlf_logger)
class DLBookModel(pl.LightningModule):
    def __init__(self):
        super(DLBookModel, self).__init__()
```

```
      . . .
  def training_step(self, batch, batch_nb):
      loss = self.log("train_loss", loss, on_epoch=True)
```

In the preceding code, we have passed an instance of `MLFlowLogger` to replace the default logger of PL. The `tracking_uri` argument controls where the logged data goes.

Other details about PyTorch integration can be found on the official website: `https://pytorch-lightning.readthedocs.io/en/stable/api/pytorch_lightning.loggers.mlflow.html`.

Setting up MLflow with DVC

To use DVC to manage large datasets, you need to install it using a package manager such as `pip`, `conda`, or `brew` (for macOS users):

```
pip install dvc
```

All the installation options can be found at `https://dvc.org/doc/install`.

Managing datasets using DVC requires a set of commands to be executed in a specific order:

1. The first step is to set up a Git repository with DVC:

    ```
    git init
    dvc init
    git commit -m 'initialize repo'
    ```

2. Now, we need to configure the remote storage for DVC:

    ```
    dvc remote add -d myremote /tmp/dvc-storage
    git commit .dvc/config -m "Added local remote storage"
    ```

3. Let's create a sample data directory and fill it with some sample data:

    ```
    mkdir data
    cp example_data.csv data/
    ```

4. At this stage, we are ready to start tracking the dataset. We just need to add our file to DVC. This operation will create an additional file, `example_data.csv.dvc`. In addition, the `example_data.csv` file gets added to `.gitignore` automatically so that Git no longer tracks the original file:

    ```
    dvc add data/example_data.csv
    ```

5. Next, you need to commit and upload the example_data.csv.dvc and .gitignore files. We will tag our first dataset as v1:

```
git add data/.gitignore data/example_data.csv.dvc
git commit -m 'data tracking'
git tag -a 'v1' -m 'test_data'
dvc push
```

6. After using the dvc push command, our data will be available on remote storage. This means we can remove the local version. To restore example_data.csv, you can simply call dvc pull:

```
dvc pull data/example_data.csv.dvc
```

7. When example_data.csv is modified, we need to add and push again to update the version on remote storage. We will tag the modified dataset as v2:

```
dvc add data/example_data.csv
git add data/example_data.csv.dvc
git commit -m 'data modification description'
git tag -a 'v2' -m 'modified test_data'
dvc push
```

After executing these commands, you will have two versions of the same dataset being tracked by Git and DVC: v1 and v2.

Next, let's look at how MLflow can be combined with DVC:

```
import mlflow
import dvc.api
import pandas as pd
data_path='data/example_data.csv'
repo='/Users/BookDL_demo/'
version='v2'
data_url=dvc.api.get_url(path=path, repo=repo, rev=version)
# this will fetch the right version of our data file
data = pd.read_csv(data_url)
# log important information using mlflow
mlflow.start_run()
mlflow.log_param("data_url", data_url)
mlflow.log_artifact(...)
```

In the preceding code snippet, `mlflow.log_artifact` was used to save information about specific columns for the experiment.

Overall, we can run multiple experiments through MLflow with different versions of the dataset tracked by DVC. Similar to W&B, MLflow also provides a web page where we can compare our experiments. All you need is to type the following command in the terminal:

```
mlflow ui
```

This command will start a web server hosting a web page on `http://127.0.0.1:5000`. The following screenshot shows the MLflow dashboard:

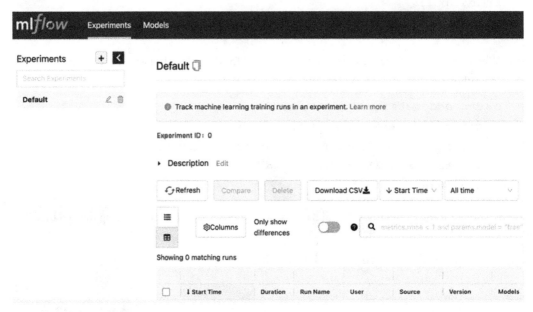

Figure 4.3 – The MLflow dashboard; new runs will be populated at the bottom of the page

Things to remember

a. MLflow can track dependencies, model parameters, metrics, and artifacts. It is often combined with DVC for efficient dataset versioning.

b. MLflow can easily be integrated with DL frameworks, including Keras, TensorFlow, and PyTorch.

c. MLflow provides an interactive visualization where multiple experiments can be analyzed at the same time.

So far, we have learned how to manage DL projects in W&B and MLflow and DVC. In the next section, we will introduce popular tools for dataset versioning.

Dataset versioning – beyond Weights & Biases, MLflow, and DVC

Throughout this chapter, we have seen how datasets can be managed by DL project-tracking tools. In the case of W&B, we can use artifacts, while in the case of MLflow and DVC, DVC runs on top of a Git repository to track different versions of datasets, thereby solving the limitations of Git.

Are there any other methods and/or tools that are useful for dataset versioning? The simple answer is yes, but again, the more precise answer depends on the context. To make the right choice, you must consider various aspects including cost, ease of use, and integration difficulty. In this section, we will mention a few tools that we believe are worth exploring if dataset versioning is one of the critical components of your project:

- **Neptune** (`https://docs.neptune.ai`) is a metadata store for MLOps. Neptune artifacts allow versioning to be conducted on datasets that are stored locally or in cloud.
- **Delta Lake** (`https://delta.io`) is an open source storage abstraction that runs on top of a data lake. Delta Lake works with Apache Spark APIs and uses distributed processing to improve throughput and efficiency.

> **Things to remember**
>
> a. There are many data versioning tools on the market. To select the right tool, you must consider various aspects including cost, ease of use, and integration difficulty.
>
> b. Tools such as W&B, MLflow, DVC, Neptune, and Delta Lake can help you with dataset versioning.

With that, we have introduced popular tools for dataset versioning. The right tool differs project by project. Therefore, you must evaluate the pros and cons of each tool before integrating one into your project.

Summary

Since DL projects involve many iterations of training models and evaluation, efficiently managing experiments, models, and datasets can help the team reach its goal faster. In this chapter, we looked at the two most popular settings for DL project tracking: W&B and MLflow integrated with DVC. Both settings provide built-in support for Keras and PL, which are the two most popular DL frameworks. We have also spent some time describing tools that put more emphasis on dataset versioning: Neptune and Delta Lake. Please keep in mind that you must evaluate each tool thoroughly to select the right tool for your project.

At this point, you are familiar with the frameworks and processes for building a proof of concept and training the necessary DL model. Starting from the next chapter, we will discuss how to scale up by moving individual components of the DL pipeline to the cloud.

Part 2 – Building a Fully Featured Product

The next phase is to migrate the proof of concept to an existing infrastructure. Throughout this process, the initial versions of the data processing logic and the models often get reimplemented using different tools and services, with the goal of increasing the throughput and improving the efficiency. In this book, we focus on AWS, the most popular web service for handling high volumes of data and expensive computations.

This part comprises the following chapters:

- *Chapter 5, Data Preparation in the Cloud*
- *Chapter 6, Efficient Model Training*
- *Chapter 7, Revealing the Secret of Deep Learning Models*

5

Data Preparation in the Cloud

In this chapter, we will learn how data preparation can be set up in the cloud by leveraging various AWS cloud services. Considering the importance of **extract, transform, and load** (ETL) operations within data preparation, we will take a deeper look into setting up and scheduling ETL jobs in a cost-efficient manner. We will cover four different setups: ETL running on a single-node EC2 instance and an EMR cluster, and then utilizing Glue and SageMaker for ETL jobs. This chapter will also introduce Apache Spark, the most popular framework for ETL. By completing this chapter, you will be able to leverage the different advantages of the presented setups and select the right set of tools for your project.

In this chapter, we're going to cover the following main topics:

- Data processing in the cloud
- Introduction to Apache Spark
- Setting up a single-node EC2 instance for ETL
- Setting up an EMR cluster for ETL
- Creating a Glue job for ETL
- Utilizing SageMaker for ETL

Technical requirements

You can download the supplemental material for this chapter from this book's GitHub repository: https://github.com/PacktPublishing/Production-Ready-Applied-Deep-Learning/tree/main/Chapter_5.

Data processing in the cloud

The success of **deep learning** (**DL**) projects depends on the quality and the quantity of data. Therefore, the systems for data preparation must be stable and scalable enough to process terabytes and petabytes of data efficiently. This often requires more than a single machine; a cluster of machines running a powerful ETL engine must be set up for the data process so that it can store and process a large amount of data.

First, we would like to introduce ETL, the core concept in data processing in the cloud. Next, we will provide an overview of a distributed system setup for data processing.

Introduction to ETL

Throughout the ETL process, data will be collected from one or more sources, get transformed into different forms as necessary, and get saved in data storage. In short, ETL itself covers the overall data processing pipeline. ETL interacts with three different types of data throughout: **structured**, **unstructured**, and **semi-structured**. While structured data represents a set of data with a schema (for example, a table), unstructured data does not have an explicit schema defined (for example, text, image, or PDF files). Semi-structured data has partial structures within the data itself (for example, HTML or emails).

Popular ETL frameworks include **Apache Hadoop** (`https://hadoop.apache.org`), **Presto** (`https://prestodb.io`), **Apache Flink** (`https://flink.apache.org`), and **Apache Spark** (`https://spark.apache.org`). Hadoop is one of the earliest data processing engines to take advantage of distributed processing. Presto is specialized in processing data in SQL, and Apache Flink is built to process streaming data. Out of these four frameworks, Apache Spark is the most popular tool as it can process every data type. *Apache Spark exploits in-memory data processing to increase its throughput* and provides much more scalable data processing solutions than Hadoop. Furthermore, it can easily be integrated with other ML and DL tools. For such reasons, we will mostly focus on Spark in this book.

Data processing system architecture

Setting up a system for data processing is not a trivial task because it involves procuring high-end machines periodically, linking various data processing software correctly, and making sure data is not lost when a failure occurs. Therefore, many companies utilize cloud services, a wide range of software services that are delivered on demand over the internet. While many companies provide various cloud services, **Amazon Web Services** (**AWS**) stands out the most with its stable and easy-to-use services.

To give you a broader picture of how complex a data processing system can be in real life, let's look at a sample system architecture based on AWS services. The *core component* of this system is open sourced *Apache Spark* carrying out the main ETL logic. A typical system also contains components for scheduling individual jobs, storing data, and visualizing the processed data:

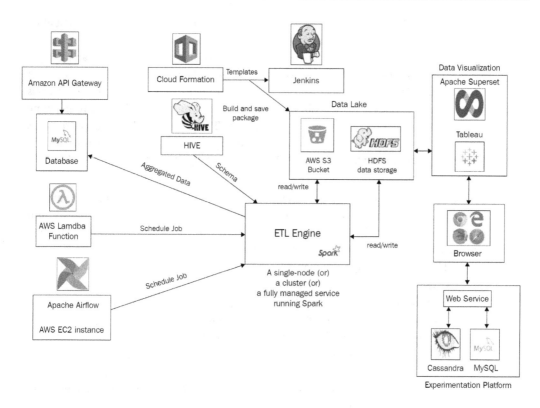

Figure 5.1 – A generic architecture for data processing pipelines
along with visualization and experimentation platforms

Let's look at each of these components:

- **Data Storage**: Data storage is responsible for keeping data and relevant metadata:

  - **Hadoop Distributed File System (HDFS)**: Open-sourced HDFS is a *distributed filesystem that can scale on demand* (`https://hadoop.apache.org`). HDFS has been the traditional pick for data storage because Apache Spark and Apache Hadoop demonstrate the best performance on HDFS.

  - **Amazon Simple Storage Service (S3)**: This is a *data storage service provided by AWS* (`https://aws.amazon.com/s3`). S3 uses the concept of objects and buckets, where an object refers to individual files and a bucket refers to a container for objects. For each project or submodule, you can create a bucket and configure the permission differently for reading and writing operations. Buckets can also apply versioning to the data, keeping track of the changes.

- **ETL Engine**: There are different ways to set up an ETL process using AWS. Each option has different advantages, and you must have an in-depth understanding of each to select the right setting for your project. You can use a single machine to keep the management simple but also utilize multiple machines or fully managed ETL services such as Amazon Glue for greater throughput:

- **Setting up a single machine for ETL: Amazon Elastic Compute** (EC2) is a virtual computing environment that's referred to as an instance or a node (`https://aws.amazon.com/ec2`). This service virtually allocates a machine of your choice: a machine with various types of CPUs/GPUs, network connections, security settings, and disks. Furthermore, there is a set of pre-configured environments called **Amazon Machine Images** (**AMIs**) running various operating systems, services, and libraries. For example, you can get an Ubuntu machine with a Jupyter notebook that supports **TensorFlow** (**TF**) projects within a couple of clicks on the web console. In the *Setting up a single-node EC2 instance for ETL* section, we will take a closer look at how to set up an EC2 instance with Spark, and a Jupyter notebook for the ETL process.

- **Setting up a cluster for ETL**: The advantage of having a cluster for ETL comes from the throughput; we can process a large amount of data more efficiently. However, there are also downsides – for example, it requires dedicated **machine learning operations** (**MLOps**) engineers for maintenance. **Elastic MapReduce** (**EMR**) *is a managed cluster platform* provided by AWS that helps set up an ETL process using multiple machines with minimal effort (`https://aws.amazon.com/emr`). Configuring EMR includes specifying the number of EC2 nodes, the type of EC2 nodes (compute intensive versus memory intensive), scripts that will run on each node, security groups, subnets, and tags. The *Setting up an EMR cluster for ETL* section is dedicated to setting up an EMR cluster for Spark-based ETL jobs.

- **Using a fully managed ETL service: Apache Glue** (`https://aws.amazon.com/glue`) *is a service designed for ETL*. Its advantage comes from the fact that it doesn't require any maintenance by MLOps. The *Creating a Glue job for ETL* section explains how to run a Spark job using Glue for ETL.

- **Utilizing an end-to-end service for ETL: SageMaker** (`https://aws.amazon.com/sagemaker`) *is an end-to-end service for ML*. You can configure SageMaker to handle data processing, model development with notebooks, model training, and deploy models to a production setting. It uses a specific set of EC2 instances that are designed for ML projects to run individual nodes. These nodes have names that start with `ml`, and costs are about 30 to 40% higher than the other EC2 instances (`https://aws.amazon.com/sagemaker/pricing`). In the *Utilizing SageMaker for ETL* section, we will describe how to set up a SageMaker for ETL process on EC2 instances.

Considering the amount of data that needs to be processed, a correctly chosen ETL service, along with an appropriate data storage selection, can improve the pipeline's efficiency significantly. The key factors to consider include the source of the data, the volume of the data, the available hardware resources, and scalability, to name a few.

- **Scheduling**: Often, ETL jobs must be periodically run (for example, daily, weekly, or monthly) and hence require a scheduler:

  - **AWS Lambda functions**: Lambda functions (`https://aws.amazon.com/lambda`) are designed to run jobs on EMR without provisioning or managing infrastructure. Execution time can be configured dynamically; the job can run right away or can be scheduled to run at different times. *The AWS Lambda function runs the code in a serverless manner so that it does not require maintenance.* If there is any error during the execution, the EMR cluster will shut down automatically.

- **Airflow**: Schedulers play an important role in automating the ETL process. Airflow (`https://airflow.apache.org`) *is one of the most popular scheduler frameworks used by data engineers.* Airflow's **Directed Acyclic Graph** (**DAG**) can be used to schedule a pipeline periodically. Airflow is more common than AWS Lambda functions for running Spark jobs periodically because Airflow makes it easy to backfill the data when any of the preceding executions failed.

- **Build**: Build is the process of deploying a code package to an AWS computing resource (such as EMR or EC2) or setting up a set of AWS services based on pre-defined specifications:

  - **CloudFormation**: CloudFormation templates (`https://aws.amazon.com/cloudformation`) *help provision cloud infrastructure as code.* CloudFormation typically does a particular task in setting up a system, such as creating an EMR cluster, preparing an S3 bucket with a particular specification, or terminating a running EMR cluster. It helps to standardize recurring tasks.

  - **Jenkins**: Jenkins (`https://www.jenkins.io`) builds executables written in Java and Scala. We use *Jenkins to build Spark pipeline artifacts (for example, .jar files) and deploy them to EMR nodes.* Jenkins also makes use of CloudFormation templates to execute a task in a standardized way.

- **Database**: The key difference between data storage and databases is that databases are used to store structured data. Here, we will discuss two popular types of databases: *relational databases* and *key-value storage databases*. We will describe how they are different and explain appropriate use cases:

  - **Relational databases**: *Relational databases store structured data with a schema in table format.* The main benefit of storing data in a structured manner comes from data management; the value of the data being stored is strictly controlled, keeping the values in a consistent format. This allows the database to make additional optimizations when storing and retrieving particular sets of data. ETL jobs generally read the data from one or more data storage services, process the data, and store the processed data in relational databases such as **MySQL** (`https://www.mysql.com`), and **PostgreSQL** (`https://www.postgresql.org`). AWS provides a relational database service as well: **Amazon RDS** (`https://aws.amazon.com/rds`).

  - **Key-value storage databases**: Unlike the traditional relational databases, these are *databases that are optimized for a high volume of read and write operations.* Such databases store data in a distinct key-value pair fashion. In general, data consists of a set of keys and a set of values that hold attributes for each key. Many of the databases support schemas, but their main advantage comes from the fact that they also support unstructured data. In other words, you can store any data, even though each of them has a different structure. Popular databases of this type include **Cassandra** (`https://cassandra.apache.org`) and **MongoDB** (`https://www.mongodb.com`). Interestingly, AWS provides a key-value storage database known as **DynamoDB** as a service (`https://aws.amazon.com/dynamodb`).

- **Metastore**: In some cases, the initial set of data that's collected and made available in data storage may not consist of any information about itself: for example, it may be missing column types or details about the source. Such information often helps engineers when they are managing and processing the data. Therefore, engineers have introduced the concept of the *metastore*, which *is a repository for metadata*. The metadata, which is stored as a table, provides the location, schema, and update history of the data it points to.

  In the case of AWS, **Glue Data Catalog** plays the role of metastore to provide built-in support for S3. Hive (`https://hive.apache.org`), on the other hand, is an open-sourced metastore for HDFS. The main advantage of Hive comes from data querying, summarization, and analysis, which comes naturally as it provides interaction based on SQL-like language.

- **Application programming interface (API) services**: *API endpoints allow data scientists and engineers to interact with the data efficiently*. For example, API endpoints can be set up to allow easy access to the data stored in the S3 bucket. Many frameworks have been designed for API services. For example, the **Flask API** (`https://flask.palletsprojects.com`) and **Django** (`https://www.djangoproject.com`) frameworks are based on Python, while the **Play** framework (`https://www.playframework.com`) is often used for projects in Scala.

- **Experimental platforms**: Evaluating system performance in production is often achieved by a popular user experience research methodology known as A/B testing. *By deploying two different versions of the system and comparing the user experiences, A/B testing allows us to understand whether the recent change have made a positive impact on the system or not*. In general, setting up A/B testing involves two components:

  - **Rest API**: *A Rest API provides greater flexibility in handling a request with different parameters and returning data in a processed manner*. Hence, it is common to set up a Rest API service that aggregates necessary data from databases or data storage for analytical purposes and provides data in JSON format to A/B experimentation platforms.

  - **A/B experimentation platform**: Data scientists often use an application with a **graphical user interface (GUI)** to schedule various A/B testing experiments and visualize the aggregated data intuitively for analysis. GrowthBook (`https://www.growthbook.io`) is an open source example of such a platform.

- **Data visualization tools**: There are a few different teams and groups within a company (for example, marketing, sales, and executives), who can benefit from intuitively visualizing the data. Data visualization tools often support custom dashboard creation, which helps with the data analysis process. Tableau (`https://www.tableau.com`) is a popular tool among project leaders, but it's proprietary software. On the other hand, Apache Superset (`https://`

`superset.apache.org`) is an open-sourced data visualization tool that supports most of the standard databases. If the management cost is a concern, Apache Superset can be configured to read and plot visualizations using data stored in serverless databases such as AWS Athena (`https://aws.amazon.com/athena`).

- **Identity Access Management (IAM)**: IAM is a permission system that regulates access to AWS resources. Through IAM, it is possible to control a set of resources that users can access and a set of operations that they can conduct on the provided resources. More details about IAM can be found at `https://aws.amazon.com/iam`.

Things to remember

a. Throughout an ETL process, data will be collected from one or more sources, transformed into different forms as necessary, and get saved into data storage or a database.

b. Apache Spark is an open source ETL engine that's widely used for processing large amounts of data of various types: structured, unstructured, and semi-structured.

c. A typical system that's been set up for a data processing job consists of various components, including a data store, databases, ETL engines, data visualization tools, and experimental platforms.

d. ETL engines can run in several settings – on a single machine, a cluster, a fully managed ETL service in the cloud, and on end-to-end services designed for DL projects.

In the next section, we will cover key programming concepts in Apache Spark, the most popular tool for ETL.

Introduction to Apache Spark

Apache Spark is an open-sourced data analytics engine that is used for data processing. The most popular use case is ETL. As an introduction to Spark, we will cover the key concepts surrounding Spark and some common Spark operations. Specifically, we will start by introducing **resilient distributed datasets** (**RDDs**) and DataFrames. Then, we will discuss Spark basics that you need to know about for ETL tasks: how to load a set of data from data storage, apply various transformations, and store the processed data. Spark applications can be implemented using multiple programming languages: Scala, Java, Python, and R. In this book, we will use Python so that we are aligned with the other implementations. The code snippets in this section can be found in this book's GitHub repository: `https://github.com/PacktPublishing/Production-Ready-Applied-Deep-Learning/tree/main/Chapter_5/spark`. The datasets we will use in our examples include Google Scholar and the COVID datasets that we crawled in *Chapter 2, Data Preparation for Deep Learning Projects*, and another COVID dataset provided by the New York Times (`https://github.com/nytimes/covid-19-data`). We will refer to the last dataset as NY Times COVID.

Resilient distributed datasets and DataFrames

The unique advantage of Spark comes from RDDs, immutable distributed collections of data objects. By exploiting RDDs, Spark can efficiently process data that exploits parallelism. The built-in parallel processing of Spark operating on RDDs helps with data processing, even when one or more of its processors fails. When a Spark job is triggered, the RDD representation of the input data gets split into multiple partitions and distributed to each node for transformations, maximizing the throughput.

Like pandas DataFrames, Spark also has the concept of DataFrames, which represent tables in a relational database with named columns. A DataFrame is also an RDD, so the operations that we describe in the next section can be applied as well. A DataFrame can be created from data structured as tables, such as CSV data, a table in Hive, or existing RDDs. DataFrames come with schemas that an RDD does not provide. As a result, an RDD is used for unstructured and semi-structured data, while a DataFrame is used for structured data.

Converting between RDDs and DataFrames

The first step for any Spark operation is to create a `SparkSession` object. Specifically, the `SparkSession` module from `pyspark.sql` is used to create a `SparkSession` object. The `getOrCreate` function from the module is used to create the session object, as shown here. A `SparkSession` object is the entry point of a Spark application. It provides a way to interact with the Spark application under different contexts, such as the Spark context, Hive context, and SQL context:

```
from pyspark.sql import SparkSession
spark_session = SparkSession.builder\
        .appName("covid_analysis")\
        .getOrCreate()
```

Converting an RDD into a DataFrame is simple. Given that an RDD does not have any schema, you can create a DataFrame without any schema, as follows:

```
# convert to df without schema
df_ri_freq = rdd_ri_freq.toDF()
```

To convert an RDD into a DataFrame with a schema, you need to use the `StructType` class, which is part of the `pyspark.sql.types` module. Once a schema has been created using the `StructType` method, the `createDataFrame` method of the Spark session object can be used to convert an RDD into a DataFrame:

```
from pyspark.sql.types import StructType, StructField,
StringType, IntegerType
# rdd for research interest frequency data
```

```
rdd_ri_freq = ...
# convert to df with schema
schema = StructType(
          [StructField("ri", StringType(), False),
           StructField("frequency", IntegerType(), False)])
df = spark.createDataFrame(rdd_ri_freq, schema)
```

Now that we have learned how to set up a Spark environment in Python, let's learn how to load a dataset as an RDD or a DataFrame.

Loading data

Spark can load data of different formats that's stored in various forms of data storage. Loading data stored in CSV format is a basic operation of Spark. This can easily be achieved using the `spark_session.read.csv` function. It reads a CSV file located locally or in the cloud, such as in an S3 bucket, as a DataFrame. In the following code snippet, we are loading Google Scholar data stored in S3. The `header` option can be used to indicate that the CSV file has a header:

```
# datasets location
google_scholar_dataset_path = "s3a://my-bucket/dataset/dataset_
csv/dataset-google-scholar/output.csv"
# load google scholar dataset
df_gs = spark_session. \
        .read \
        .option("header", True) \
        .csv(google_scholar_dataset_path)
```

The following figure shows the results of `df_gs.show(n=3)`. The `show` function prints the first *n* rows, along with the column headings:

```
+---------------+----------+--------------------+--------------------+--------------------+
|    author_name|     email|         affiliation|    coauthors_names|   research_interest|
+---------------+----------+--------------------+--------------------+--------------------+
| William Eberle|tntech.edu|Tennessee Technol...|                null|data_mining##anom...|
|Lawrence Holder|   wsu.edu|Washington State ...|Diane J Cook##Wil...|artificial_intell...|
|     Talbert DA|tntech.edu|Tennessee Technol...|                null|machine_learning#...|
+---------------+----------+--------------------+--------------------+--------------------+
only showing top 3 rows
```

Figure 5.2 – A sample DataFrame created by loading a CSV file

Similarly, a JSON file from data storage can be read using the `read.json` function of the `SparkSession` module:

```
# loada json file
json_file_path="s3a://my-bucket/json/cities.json"
df = spark_session.read.json(json_file_path)
```

In the next section, we will learn how to process loaded data using Spark operations.

Processing data using Spark operations

Spark provides a set of operations that transforms an RDD into an RDD of a different structure. Implementing a Spark application is the process of chaining a set of Spark operations on an RDD to transform the data into the target format. In this section, we will discuss the most commonly used – that is, `filter`, `map`, `flatMap`, `reduceByKey`, `take`, `groupBy`, and `join`.

filter

In most cases, filters are often applied first to drop unnecessary data. Applying the `filter` method to a DataFrame can help you choose the rows of interest from the given DataFrame. In the following code snippet, we are using this method to only keep the rows where `research_interest` is not None:

```
# research_interest cannot be None
df_gs_clean = df_gs.filter("research_interest != 'None'")
```

map

Like the `map` function in other programming languages, the `map` operation in Spark applies the given function to each data entry. Here, we are using the `map` function to only keep the `research_interest` column:

```
# we are only interested in research_interest column
rdd_ri = df_gs_clean.rdd.map(lambda x: (x["research_
interest"]))
```

flatMap

The `flatMap` function flattens the RDD after applying the given function to every entry and returns the new RDD. In this example, the `flatMap` operation splits each data entry with the ## separator and then creates a pair of `research_interest` and a default frequency with a value of 1:

```
# raw research_interest data into pairs of research area and a
frequency count
```

```
rdd_flattened_ri = rdd_ri.flatMap(lambda x: [(w.lower(), 1) for
w in x.split('##')])
```

reduceByKey

reduceByKey groups the input RDD based on its key. Here, we are using `reduceByKey` to sum the frequencies to understand the number of occurrences for each `research_interest`:

```
# The pairs are grouped based on research area and the
frequencies are summed up
rdd_ri_freq = rdd_flattened_ri.reduceByKey(add)
```

take

One of the basic operations of Spark is `take`. This function is used to get the first *n* elements from an RDD:

```
# we are interested in the first 5 entries
rdd_ri_freq_5 = rdd_ri_freq.take(5)
```

Grouping operations

The idea of grouping is to collect identical data entries within a DataFrame into groups and perform aggregation (for example, average or summation) on the groups.

As an example, let's employ the Moderna COVID dataset to get the average number of doses allocated per jurisdiction (state) using the `groupby` operation. Here, we are using the `sort` function to sort the state-wise average number of doses. The `toDF` and `alias` functions can help add a name for the new DataFrame:

```
# calculate average number of 1st corona vaccine per
jurisdiction (state)
df_avg_1 = df_covid.groupby("jurisdiction")\
  .agg(F.avg("_1st_dose_allocations")
  .alias("avg"))\
  .sort(F.col("avg").desc())\
  .toDF("state", "avg")
```

While applying `groupby`, multiple aggregations (sum and avg) can be applied in a single command. The columns that get created from aggregated functions such as `F.avg` or `F.sum` can be renamed using `alias`. In the following example, aggregations are being performed on the Moderna COVID dataset to get the average number and sum of the first and second doses:

```
# At jurisdiction (state) level, calculate at average weekly
1st & 2nd dose vaccine distribution. Similarly calculate sum
```

```
for 1st and 2nd dose
df_avg = df_covid.groupby(F.lower("jurisdiction").
alias("state"))\
    .agg(F.avg("_1st_dose_allocations").alias("avg_1"), \
        F.avg("_2nd_dose_allocations").alias("avg_2"), \
        F.sum("_1st_dose_allocations").alias("sum_1"), \
        F.sum("_2nd_dose_allocations").alias("sum_2")
        ) \
    .sort(F.col("avg_1").desc())
```

The calculation is performed at the state level using the groupby function. This dataset contains 63 states in total, including certain entities (federal agencies) as a state.

join

The join functionality helps combine rows from two DataFrames.

To demonstrate how join can be used, we will join the Moderna COVID dataset with the NY Times COVID dataset. Before we explain any join operations, we must apply aggregation to the NY Times COVID dataset, just like how we processed the Moderna COVID dataset previously. In the following code snippet, the groupby operation is being applied at the state level to get the aggregated (sum) value representing the total number of deaths and the total number of cases:

```
# at jurisdiction (state) level, calculate total number of
deaths and total number of cases
df_cases = df_covid2 \
        .groupby(F.lower("state").alias("state")) \
        .agg(F.sum("deaths").alias("sum_deaths"), \
            F.sum("cases").alias("sum_cases"))
```

Figure 5.3 shows the results of the df_cases.show(n=3) operation, which visualizes the top three rows of the processed DataFrame:

```
+-------------+----------+------------+
|        state|sum_deaths|   sum_cases|
+-------------+----------+------------+
|west virginia| 1286901.0|   7.631901E7|
|new hampshire|  620816.0|  4.3191729E7|
|      alabama| 5005646.0|2.68440532E8|
+-------------+----------+------------+
```

Figure 5.3 – The top three rows of the aggregated results

We are now ready to demonstrate the two types of join: equi-join and left join.

Equi-join (inner-join)

Equi-join, also called an inner-join, is the default `join` operation in Spark. An inner join is used to join two DataFrames on common column values. The rows where the keys don't match will get dropped in the final DataFrame. In this example, equi-join will be applied to the `state` column as a common column between the Moderna COVID dataset and the NY Times COVID dataset.

The first step is to create aliases for the DataFrames using `alias`. Then, we call the `join` function on one DataFrame while passing the other DataFrame that defines the column relationship and the type of join:

```
# creating an alias for each DataFrame
df_moderna = df_avg.alias("df_moderna")
df_ny = df_cases.alias("df_ny")

df_inner = df_moderna.join(df_ny, F.col("df_moderna.state") ==
F.col("df_ny.state"), 'inner')
```

The following is the output of the `df_inner.show(n=3)` operation:

```
+-------------------+-------------------+----------+----------+-----------+-----------+
|state              |state              |avg_1     |avg_2     |sum_deaths |sum_cases  |
+-------------------+-------------------+----------+----------+-----------+-----------+
|west virginia      |west virginia      |27675.0   |27675.0   |1286901.0  |7.631901E7 |
|new hampshire      |new hampshire      |20711.25  |20711.25  |620816.0   |4.3191729E7|
|alabama            |alabama            |70745.625 |70745.625 |5005646.0  |2.68440532E8|
+-------------------+-------------------+----------+----------+-----------+-----------+
```

Figure 5.4 – The output of using the df_inner.show(n=3) operation

Now, let's look at the other type of join, left join.

Left join

A left join is another popular `join` operation for data analysis. A left join returns all the rows from one DataFrame, regardless of the matches found on the other DataFrame. When the `join` expression does not match, it assigns `null` for the missing entries.

The left join syntax is like that of equi-join. The only difference is that you need to pass the `left` keyword when specifying the join type instead of `inner`. The left join takes all the values of the mentioned column (`df_m.state`) in the first DataFrame mentioned (`df_m`). Then, it tries to match entries with the DataFrame mentioned second (`df_ny`) on the column mentioned (`df_ny.state`). In this example, if a particular state appears on both DataFrames, the output of the `join` operation will be the state, along with values from both DataFrames. If a particular state is only available in the first DataFrame (`df_m`) but not in the second (`df_ny`), then it will add the state with the values for the first DataFrame only, keeping the other entry as `null`:

```
# join results in all rows from the left table. Missing entries
from the right table will result in "null"
```

```
df_left = df_moderna.join(df_ny, F.col("df_m.state") ==
F.col("df_ny.state"), 'left')
```

The output of df_left.show(n=3) is shown here:

```
+--------------------+--------+--------+--------+--------+--------------+-----------+-----------+
|               state|   avg_1|   avg_2|   sum_1|   sum_2|         state|sum_deaths|  sum_cases|
+--------------------+--------+--------+--------+--------+--------------+-----------+-----------+
|       west virginia| 27675.0| 27675.0|442800.0|442800.0|  west virginia| 1286901.0|  7.631901E7|
|       new hampshire|20711.25|20711.25|331380.0|331380.0|  new hampshire|  620816.0|4.3191729E7|
|      mariana islands|   780.0|     0.0| 11700.0|     0.0|          null|      null|       null|
```

Figure 5.5 – The output of the df_inner.show(n=3) operation

Even though Spark provides a wide range of operations that cover vastly different cases, you may find building a custom operation more useful due to the complexity of your logic.

Processing data using user-defined functions

A **user-defined function** (**UDF**) *is a reusable custom function that performs a transformation on an RDD.* A UDF function can be reused on several DataFrames. In this section, we will provide a complete code example for processing the Google Scholar dataset using UDF.

First of all, we would like to introduce the pyspark.sql.function module, which allows you to define a UDF with the udf method and provides various column-wise operations. pyspark.sql.function also includes functions for aggregations such as avg or sum for computing the average and total, respectively:

```
import pyspark.sql.functions as F
```

In the Google Scholar dataset, data_science, artificial_intelligence, and machine_learning all refer to the same field of **artificial intelligence** (**AI**). So, it would be good to create a UDF for cleaning up this field. First, it will take an RDD of the research_interest data and check if any of the data can be categorized as AI. If matches are found, it puts a value of 1 in a new column. It will assign 0 otherwise. The results of the UDF are stored in a new column called is_artificial_intelligence using the withColumn method. In the following code snippet, the @F.udf annotation informs Spark that the function is a UDF. The col method from pyspark.sql.functions is often used to pass a column as an argument for UDF. Here, F.col("research_interest") has been passed to the UDF is_ai method, indicating which column that UDF should operate on:

```
# list of research_interests that are under same domain
lst_ai  = ["data_science", "artificial_intelligence",
           "machine_learning"]
@F.udf
```

```
def is_ai(research):
    """ return 1 if research in AI domain else 0"""
    try:
        # split the research interest with delimiter "##"
        lst_research = [w.lower() for w in str(research).
split("##")]
        for res in lst_research:
            # if present in AI domain
            if res in lst_ai:
                return 1
        # not present in AI domain
        return 0
    except:
        return -1
# create a new column "is_artificial_intelligence"
df_gs_new = df_gs.withColumn("is_artificial_intelligence",\ is_
ai(F.col("research_interest")))
```

After processing the raw data, we want to store it in data storage so that we can reuse it for other purposes.

Exporting data

In this section, we will learn how to save a DataFrame into an S3 bucket. In the case of RDD, it must be converted into a DataFrame to be saved appropriately.

Typically, data analysts want to write the aggregated data as a CSV file for the following operations. To export a DataFrame as a CSV file, you must use the df.write.csv function. In the case of text values, we recommend that you use option("quoteAll", True), which will encapsulate each value with quotes.

In the following example, we are providing an S3 path to generate a CSV file in an S3 bucket. coalesce(1) is used to write a single CSV file instead of multiple CSV files:

```
S3_output_path = "s3a:\\my-bucket\output\vaccine_state_avg.csv"
# writing a DataFrame as a CSVfile
sample_data_frame.\
        .coalesce(1) \
        .write \
```

```
        .mode ("overwrite") \
        .option ("header", True) \
        .option ("quoteAll", True) \
        .csv (s3_output_path)
```

If you want to save the DataFrame as a JSON file, you can use `write.json`:

```
S3_output_path = "s3a:\\my-bucket\output\vaccine_state_avg.
json"
# Writing a DataFrame as a json file
sample_data_frame \
        .write \
        .json (s3_output_path)
```

At this point, you should see that a file is stored in the S3 bucket.

Things to remember

a. An RDD is an immutable distributed collection of sets that gets split into multiple partitions and computed in different nodes of a cluster.

b. A Spark DataFrame is equivalent to a table in a relational database with named columns.

c. Spark provides a set of operations that transforms an RDD into an RDD that has a different structure. Implementing a Spark application is the process of chaining a set of Spark operations on an RDD to transform the data into the target format. You can build a custom Spark operation using UDF.

In this section, we described the basics of Apache Spark, which is the most common tool for ETL. Starting from the next section, we will talk about how to set up a Spark job in the cloud for ETL. First, let's look at how to run ETL on a single EC2 instance.

Setting up a single-node EC2 instance for ETL

EC2 instances can have various combinations of CPU/GPU, memory, storage, and network capacity. You can find configurable options for EC2 in the official documentation: `https://aws.amazon.com/ec2/instance-types`.

When creating an EC2 instance, you can choose a Docker image to run which has been predefined for various projects. These are called **Amazon Machine Images** (**AMIs**). For example, there's an image with TF version 2 installed for DL projects and an image with Anaconda set up for generic ML projects, as shown in the following screenshot. For the complete list of AMIs, please refer to `https://docs.aws.amazon.com/AWSEC2/latest/UserGuide/AMIs.html`:

Figure 5.6 – Selecting an AMI for an EC2 instance

AWS offers **Deep Learning AMIs** (**DLAMIs**), which are AMIs that are created for DL projects; images utilize different CPU and GPU configurations and different compute architectures (`https://docs.aws.amazon.com/dlami/latest/devguide/options.html`).

As mentioned in *Chapter 1*, *Effective Planning of Deep Learning-Driven Projects*, many data scientists make use of EC2 instances to develop their algorithms, exploiting the flexibility in dynamic resource allocation. The steps for creating an EC2 instance and installing Spark are as follows:

1. Create a **Virtual Private Network** (**VPN**) to restrict access to the EC2 instance for security purposes.
2. Create a `.pem` key with an EC2 key pair. A `.pem` file is used to perform authentication when a user attempts to log into the EC2 instance from a terminal.
3. Create an EC2 instance from a Docker image with the necessary tools and packages.
4. Add an inbound rule that enables access to the new instances from your local terminal.
5. Use SSH to access the EC2 instance with the `.pem` file that was created in *Step 2*.
6. Initiate the Spark shell.

We have included detailed descriptions for each step, along with screenshots, at `https://github.com/PacktPublishing/Production-Ready-Applied-Deep-Learning/tree/main/Chapter_5/ec2`.

> **Things to remember**
>
> a. An EC2 instance can have various combinations of CPU/GPU, memory, storage, and network capacity
>
> b. An EC2 instance can be created from a predefined Docker image (AMI) with a couple of clicks on the AWS web console

Next, we will learn how to set up a cluster that runs a set of Spark workers as a group.

Setting up an EMR cluster for ETL

In the case of DL, the computational power of a single EC2 instance may not be sufficient for model training or data processing. Therefore, a group of EC2 instances is often put together to increase the throughput. AWS has a dedicated service for this purpose: **Amazon Elastic MapReduce (EMR)**. It is a fully managed cluster platform that provides distributed systems for big data frameworks such as Apache Spark and Hadoop. In general, an EMR cluster that's been set up for ETL reads data from AWS storage (Amazon S3), processes the data, and writes it back to AWS storage. Spark jobs are often used to handle the ETL logic that interacts with S3. EMR provides an interesting feature named **Workspace** that helps organize notebooks by developers and shares them with other EMR users for collaborative work.

A typical EMR setup contains a master node and a few core nodes. In the case of a multi-node cluster, there must be at least one core node. A master node manages a cluster that runs the distributed application (for example, Spark or Hadoop). Core nodes are managed by the master node and run data processing tasks and store data in data storage (for example, S3 or HDFS).

Task nodes are managed by the master node and are optional. They increase the throughput of the distributed application running on the cluster by introducing another parallelism during computation. They run data processing tasks but do not store data in data storage.

The following screenshot shows the EMR cluster creation page. Throughout the form, we need to provide the cluster's name, launch mode, EMR release, applications (for example, Apache Spark for data processing and Jupyter for notebooks) to run on the cluster, and specifications of the EC2 instances. Data processing with DL often needs instances of high computational power. In the other cases, you can construct a cluster with increased memory limits:

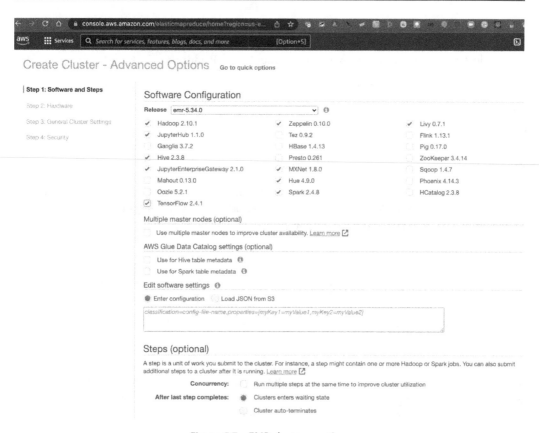

Figure 5.7 – EMR cluster creation

The detailed steps are as follows:

- **Step 1: Software and Steps**: Here, you must choose the software-related configuration – that is, the EMR release and applications (Spark, JupyterHub, and so on).

- **Step 2: Hardware**: Here, you must choose the hardware-related configuration – that is, the instance type, number of instances, and the VPN network.

- **Step 3: General Cluster Setting**: Choose the cluster name and the S3 bucket path for operational logs.

- **Step 4: Security**: You need to configure the security group and the .pem file:

  - **Security Group**: A security group needs to be chosen for the master/core nodes in the EMR cluster. A security group explains who can or can't access the nodes in the EMR. A master security group is the security group that's applied to the master node of an EMR cluster. In the master security group, you need to add a new inbound rule for the Jupyter notebook (port 9942) and open access for your IP address. If your IP address is 203.0.113.25, then you should add 203.0.113.25/32. You can also provide an IP address of 0.0.0.0/0, but you should be cautious since Jupyter applications, like other applications running in the cluster, can be accessed unexpectedly through any IP address.

- **.pem file**: A new .pem file is only needed if you want to log in to the EC2 master node and work on the Spark shell as in the case of a single EC2 instance.

After following these steps, you will need to wait for a few minutes until the state of the cluster changes to `running`. Then, you can navigate to the endpoint provided by the EMR cluster to open a Jupyter notebook. The username is `jovyan` and the password is `jupyter`.

Our GitHub repository provides step-by-step instructions for this process, along with screenshots (`https://github.com/PacktPublishing/Production-Ready-Applied-Deep-Learning/tree/main/Chapter_5/emr`).

> **Things to remember**
>
> a. EMR is a fully managed cluster platform that runs big data ETL frameworks such as Apache Spark
>
> b. You can create an EMR cluster with various EC2 instances through the AWS web console

The downside of EMR comes from the fact that it needs to be managed explicitly. An organization often has a group of developers dedicated to handling issues related to EMR clusters. Unfortunately, this can be a difficult thing to do if the organization is small. In the next section, we will introduce Glue, which doesn't require any explicit cluster management.

Creating a Glue job for ETL

AWS Glue (`https://aws.amazon.com/glue`) supports data processing in a serverless fashion. The computational resource of Glue is managed by AWS, so less effort is needed for maintenance, unlike in the case of dedicated clusters (for example, EMR). Other than the minimal maintenance effort for the resources, Glue provides additional features such as a built-in scheduler and Glue Data Catalog, which will be discussed later.

First, let's learn how to set up data processing jobs using Glue. Before you start defining the logic for data processing, you must create a Glue Data Catalog that contains the schema for the data in S3. Once a Glue Data Catalog has been defined for the input data, you can use the Glue Python editor to define the details of the data processing logic (*Figure 5.8*). The editor provides a basic setup for your application to reduce the difficulties in setting up a Glue job: `https://docs.aws.amazon.com/glue/latest/dg/edit-script.html`. On top of this template code, you will read in the Glue Data Catalog as an input, process it, and store the processed output. Since Glue Data Catalog has a nice integration for Spark, the operations within a Glue job are often achieved using Spark:

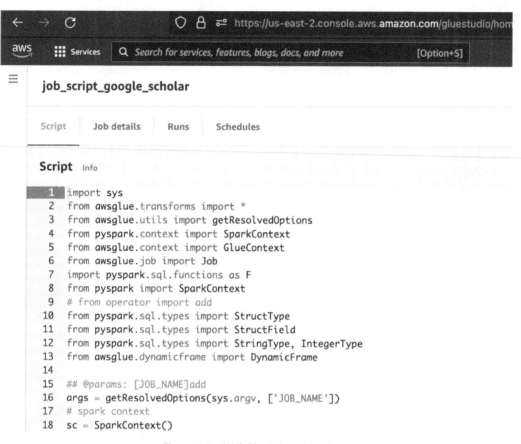

```
job_script_google_scholar

Script    Job details    Runs    Schedules

Script  Info

1  import sys
2  from awsglue.transforms import *
3  from awsglue.utils import getResolvedOptions
4  from pyspark.context import SparkContext
5  from awsglue.context import GlueContext
6  from awsglue.job import Job
7  import pyspark.sql.functions as F
8  from pyspark import SparkContext
9  # from operator import add
10 from pyspark.sql.types import StructType
11 from pyspark.sql.types import StructField
12 from pyspark.sql.types import StringType, IntegerType
13 from awsglue.dynamicframe import DynamicFrame
14
15 ## @params: [JOB_NAME]add
16 args = getResolvedOptions(sys.argv, ['JOB_NAME'])
17 # spark context
18 sc = SparkContext()
```

Figure 5.8 – AWS Glue job script editor

In the following sections, you will learn how to set up a Glue job using the Google Scholar dataset, which is stored in an S3 bucket. The complete implementation can be found at https://github. com/PacktPublishing/Production-Ready-Applied-Deep-Learning/tree/ main/Chapter_5/glue.

Creating a Glue Data Catalog

First, we will create a Glue Data Catalog (see *Figure 5.9*). Glue can only read a set of data where the metadata is stored in the Glue Data Catalog. Data Catalog consists of databases, which are collections of metadata in the form of a table. Glue provides a feature called a **crawler**, which *creates metadata for the data files present in data storage* (for example, an S3 bucket):

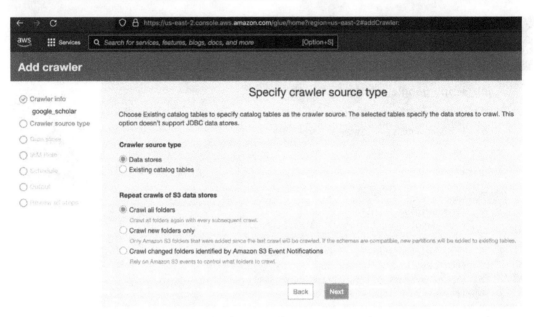

Figure 5.9 – The first step of setting up a crawler

The preceding screenshot shows the first step of creating a crawler. Details of each step can be found at https://docs.aws.amazon.com/glue/latest/dg/add-crawler.html.

Setting up a Glue context

If you look at the template code provided by AWS for Glue, you will find that some key packages are already imported. getResolvedOptions from the awsglue.utils module helps utilize the arguments that are passed to the Glue script during runtime:

```
from awsglue.utils import getResolvedOptions
args = getResolvedOptions(sys.argv, ['JOB_NAME'])
```

For a Glue job with Spark, a Spark context must be created and passed to GlueContext. A Spark session object can be accessed from a Glue context. A Glue job can be instantiated using the awsglue.job module by passing a Glue context object:

```
from pyspark.context import SparkContext
from awsglue.context import GlueContext
from awsglue.job import Job
# glue_job_google_scholar.py
# spark context
spark_context = SparkContext()
```

```
# glue context
glueContext = GlueContext(spark_context)
# spark
spark_session = glueContext.spark_session
# job
job = Job(glueContext)
# initialize job
job.init(args['JOB_NAME'], args)
```

Next, we will learn how to read data from Glue Data Catalog.

Reading data

In this section, you will learn how to read data located in an S3 bucket within the Glue context after creating a Glue table catalog.

The data in Glue passes from transform to transform using a specific data structure called a DynamicFrame, which is an extension of an Apache Spark DataFrame. DynamicFrame, with its self-describing nature, does not require any schema. This additional property of a DynamicFrame helps accommodate the data that does not conform to a fixed schema, unlike in Spark DataFrames. The required library can be imported from `awsglue.dynamicframe`. This package makes converting a DynamicFrame into a Spark DataFrame easy:

```
from awsglue.dynamicframe import DynamicFrame
```

In the following example, we are creating a Glue Data Catalog table named `google_authors` in a database named `google_scholar`. Once the database is available, `glueContext.create_dynamic_frame.from_catalog` can be used to read the `google_authors` table in the `google_scholar` database and load it as a Glue DynamicFrame:

```
# glue context
google_authors = glueContext.create_dynamic_frame.from_catalog(
            database="google_scholar",
            table_name="google_authors")
```

A Glue DynamicFrame can be converted into a Spark DataFrame using the `toDF` method. This conversion is required to apply Spark operations to the data:

```
# convert the glue DynamicFrame to Spark DataFrame
google_authors_df = google_authors.toDF()
```

Now, let's define the data processing logic.

Defining the data processing logic

Basic transformations that you can perform on a Glue DynamicFrame are provided by the `awsglue.transforms` module. These transformations include `join`, `filter`, `map`, and many others (`https://docs.aws.amazon.com/glue/latest/dg/built-in-transforms.html`). You can use them similarly to what was presented in the *Introduction to Apache Spark* section:

```
from awsglue.transforms import *
```

Additionally, every Spark operation described in the *Processing data using Spark operations* section can be applied to data in Glue if the Glue DynamicFrame has already been converted into a Spark DataFrame.

Writing data

In this section, we will learn how to write data in Glue DynamicFrame to an S3 bucket.

Given a Glue DynamicFrame, you can store the data in the given S3 path using `write_dynamic_frame.from_options` of a Glue context. You need to call the `commit` method of a job at the end to perform individual operations:

```
# path for output file
path_s3_write= "s3://google-scholar-csv/write/"
# write to s3 as a CSV file with separator |
glueContext.write_dynamic_frame.from_options(
    frame = dynamic_frame_write,
    connection_type = "s3",
    connection_options = {
            "path": path_s3_write
                        },
    format = "csv",
    format_options={
            "quoteChar": -1,
            "separator": "|"
                })
job.commit()
```

In the case of a Spark DataFrame, you must convert it into a DynamicFrame before you can store the data. The `DynamicFrame.fromDF` function takes in a Spark DataFrame object, a Glue context object, and the name of the new DynamicFrame:

```
# create a DynamicFrame from a Spark DataFrame
dynamic_frame = DynamicFrame.fromDF(df_sort, glueContext,
"dynamic_frame")
```

Now, you can use both Spark operations and Glue transformations to process your data.

> **Things to remember**
>
> a. AWS Glue is a fully managed service designed for ETL operations
>
> b. AWS Glue is a serverless architecture, which means the underlying servers will be maintained by AWS
>
> c. AWS Glue provides a built-in editor with Python boilerplate code. In this editor, you can define your ETL logic and also leverage Spark

As the last setting for ETL, we will look at SageMaker.

Utilizing SageMaker for ETL

In this section, we will describe how to set up an ETL process using SageMaker (the following screenshot shows the web console for SageMaker). *The main advantage of SageMaker comes from the fact that it is a fully managed infrastructure for building, training, and deploying ML models.* The downside is the fact that it is more expensive than EMR and Glue.

SageMaker Studio is a web-based development environment for SageMaker. SageMaker has been introduced with the philosophy that it's an all-in-one place for a data analytics pipeline. Every phase of an ML pipeline can be achieved using SageMaker Studio: data processing, algorithm design, scheduling jobs, experiment management, developing and training models, creating inference endpoints, detecting data drift, and visualizing model performance. SageMaker Studio notebooks can also be connected to EMR for computations with some restrictions; only limited Docker images (such as `Data Science` or `SparkMagic`) can be used (`https://docs.aws.amazon.com/sagemaker/latest/dg/studio-notebooks-emr-cluster.html`):

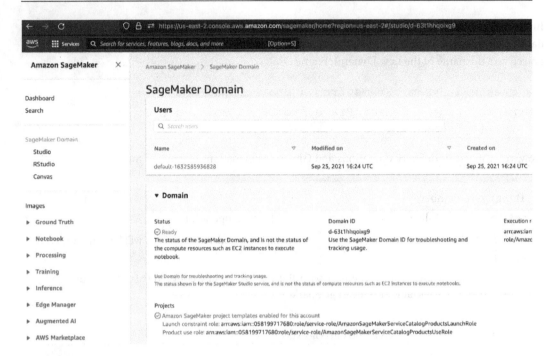

Figure 5.10 – The SageMaker web console

SageMaker provides various predefined development environments as Docker images. Popular environments are those for DL projects that have PyTorch, TF, and Anaconda installed already. A notebook can easily be attached to any of these images from the web-based development environment, as shown in the following screenshot:

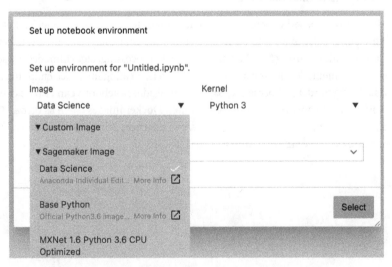

Figure 5.11 – Updating the development environment dynamically for a SageMaker notebook

The process of creating an ETL job can be broken down into four steps:

1. Create a user within SageMaker Studio.
2. Create a notebook under the user by selecting the right Docker image.
3. Define data processing logic.
4. Schedule a job.

Steps 1 and *2* are one click away in the SageMaker web console. *Step 3* can be set up using Spark. To schedule a job (*Step 4*), first, you need to install the `run-notebook` command-line utility via the `pip` command:

```
pip install https://github.com/aws-samples/sagemaker-run-
notebook/releases/download/v0.20.0/sagemaker_run_notebook-
0.20.0.tar.gz
```

Before looking at the `run-notebook` command for scheduling a notebook, we will briefly discuss the `cron` command, which defines the format for a schedule. As shown in the following diagram, six numbers are used to represent a timestamp. For example, `45 22 ** 6*` represents a schedule for 10:45 P.M. every Saturday. The * (asterisk) wildcard represents every value of the corresponding unit:

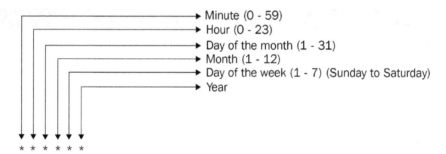

Figure 5.12 – Cron schedule format

The `run-notebook` command takes in a schedule represented with `cron` and a notebook. In the following example, `notebook.ipynb` has been scheduled to run at 8 A.M. every day in 2021:

```
run-notebook schedule --at "cron(0 8 * * * 2021)" --name
nightly notebook.ipynb
```

We have provided a set of screenshots for each step in our GitHub repository: `https://github.com/PacktPublishing/Production-Ready-Applied-Deep-Learning/blob/main/Chapter_5/sagemaker/sagemaker_studio.md`.

In the remaining sections, we will take a deeper look at how to utilize the SageMaker notebook to run a data processing job.

Creating a SageMaker notebook

A notebook instance is an ML compute instance that runs the Jupyter notebook application. SageMaker will create this instance, along with the associated resources. The Jupyter notebook is used to process data, train models, and deploy and validate the model. A notebook instance can be created in a few steps. The complete description can be found at `https://docs.aws.amazon.com/sagemaker/latest/dg/howitworks-create-ws.html`:

1. Go to the SageMaker web console: `https://console.aws.amazon.com/sagemaker`. Please note that you will need to log in with AWS credentials.

2. Under **Notebook instances**, choose **Create notebook instance**.

3. On the **Create notebook instance** page, provide the notebook instance's name and instance type. Additionally, a shell script can be configured to run when the instance is started – that is, a life cycle configuration script (see *Figure 5.13*). For example, you may want to install a set of dependency libraries (such as `pip install tensorflow`) on each new notebook. Various examples of this can be found at `https://github.com/aws-samples/amazon-sagemaker-notebook-instance-lifecycle-config-samples/tree/master/scripts`:

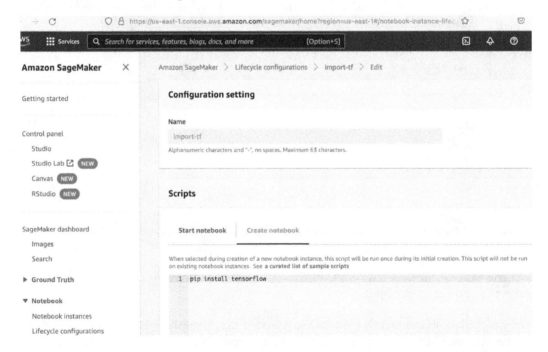

Figure 5.13 – Life cycle configuration script for a SageMaker notebook

While running a set of operations directly from the SageMaker notebook is an option, the SageMaker notebook supports running a data processing job defined explicitly outside of the notebook to increase throughput and reusability. Let's look at how we can run a Spark job from a notebook.

Running a Spark job through a SageMaker notebook

Once a notebook is ready, you can configure a Spark job using the sagemaker.processing module and execute it using a set of computational resources. SageMaker provides the PySparkProcessor class, which provides a handle for the Spark job (https://sagemaker.readthedocs.io/en/ stable/amazon_sagemaker_processing.html#data-processing-with-spark). Its constructor takes in basic setup details, such as the job's name and Python version. It takes in three parameters – framework_version, py_version, and container_version – which are used to pin the pre-built Spark containers to run the processing job. A custom image can be registered and made available on the **Elastic Container Registry** (**ECR**), which provides a secure, scalable, and reliable registry for Docker images (https://aws.amazon.com/ecr). You can choose a custom image to run in your container if you pass the ECR image URL to the image_uri parameter. image_uri will override the framework_version, py_version, and container_version parameters:

```
From sagemaker.processing import PySparkProcessor,
ProcessingInput
# ecr image URI
ecr_image_uri = '664544806723.dkr.ecr.eu-central-1.amazonaws.
com/linear-learner:latest'
# create PySparkProcessor instance with initial job setup
spark_processor = PySparkProcessor(
    base_job_name="my-sparkjob", # job name
    framework_version="2.4", # tensorflow version
    py_version="py37", # python version
    container_version="1", # container version
    role="myiamrole", # IAM role
    instance_count=2, # ec2 instance count
    instance_type="ml.c5.xlarge", # ec2 instance type
    max_runtime_in_seconds=1200, # maximum run time
    image_uri=ecr_image_uri # ECR image
)
```

In the preceding code, a PySparkProcessor class has been used to create a Spark instance. It takes in base_job_name (job name: my-sparkjob), framework_version (the TensorFlow framework version: 2.0), py_version (the Python version: py37), container_version (the container version: 1), role (the IAM role for SageMaker: myiamrole), instance_count (the number of EC2 instances: 2), instance_type (the EC2 instance type: ml.c5.xlarge), max_runtime_in_second (the maximum runtime in seconds before timeout: 1200), and image_url (the URL of the Docker image: ecr_image_uri).

Next, we will discuss the run method of PySparkProcessor, which starts the provided script through Spark:

```
# input s3 path
path_input = "s3://mybucket/input/"
# output s3 path
path_output = "s3://mybucket/output/"
# run method to execute job
spark_processor.run(
    submit_app="process.py", # processing python script
    arguments=['input', path_input, # input argument for script
                'output', path_output # input argument for
script
                ])
```

In the preceding code, the run method of PySparkProcessor executes the given script, along with the arguments provided. It takes in submit_app (a data processing job written in Python) and arguments. In this example, we have defined where the input data is located and where the output should be stored.

Running a job from a custom container through a SageMaker notebook

In this section, we will discuss how to run a data processing job from a custom image. SageMaker provides the Processor class as part of the sagemaker.processing module for this purpose. In this example, we will use the ProcessingInput and ProcessingOutput classes to create input and output objects, respectively. These objects will be passed to the run method of the Processor instance. The run method executes the data processing job:

```
# ecr image URI
ecr_image_uri = '664544806723.dkr.ecr.eu-central-1.amazonaws.
com/linear-learner:latest'
# input data path
path_data = '/opt/ml/processing/input_data'
# output data path
path_data = '/opt/ml/processing/processed_data'
```

```
# s3 path for source
path_source = 's3://mybucket/input'
# s3 path for destination
path_dest = 's3://mybucket/output'
# create Processor instance
processor = Processor(image_uri=ecr_image_uri, # ECR image
            role='myiamrole', # IAM role
            instance_count=1, # instance count
            instance_type="ml.m5.xlarge" # instance type
        )
# calling "run" method of Processor instance
processor.run(inputs=[ProcessingInput(
            source=path_source, # input source
            destination=path_data # input destination)],
        outputs=[ProcessingOutput(
            source=path_data, # output source
            destination=path_dest # output destination)],
))
```

In the preceding code, first, we create a `Processor` instance. It takes in `image_uri` (the ECR image's URL path: `ecr_image_uri`), `role` (the IAM role that has access to the ECR image: `myiamrole`), `instance_count` (the EC2 instance count: 1), and `instance_type` (the EC2 instance type: `ml.m5.xlarge`). The `run` method of the `Processor` instance can execute the job. It takes in `inputs` (the input data passed as a `ProcessingInput` object) and `outputs` (the output data passed as a `ProcessingOutput` object). While `Processor` provides a similar set of methods to `PySparkProcessor`, the main difference comes from what the `run` function takes in; `PySparkProcessor` takes in a Python script that runs Spark operations, while `Processor` takes in a Docker image that supports various types of data processing jobs.

For those who are willing to dig into the details, we recommend reading `https://docs.aws.amazon.com/sagemaker/latest/dg/build-your-own-processing-container.html`.

Things to remember

a. SageMaker is a fully managed infrastructure for building, training, and deploying ML models.

b. SageMaker provides a set of predefined development environments that users can change dynamically based on their needs.

c. SageMaker notebooks support data processing jobs defined outside of the notebook through the `sagemaker.processing` module.

Having gone through the four most popular ETL tools in AWS, let's compare the four options side by side.

Comparing the ETL solutions in AWS

So far, we have looked at four different ways of setting up ETL pipelines using AWS. In this section, we will summarize the four setups in a single table (*Table 5.1*). Some of the comparison points include support for serverless architecture, the availability of a built-in scheduler, and variety in terms of the supported EC2 instance types.

Supports	Single-Node EC2 Instance	Glue	EMR	SageMaker
Support for serverless architecture	No	Yes	No	No
Availability of a built-in workspace for collaboration among developers	No	No	Yes	No
Variety of EC2 instance types	More	Less	More	More
Availability of a built-in scheduler	No	Yes	No	Yes
Availability of a built-in job monitoring UI	No	Yes	No	Yes
Availability of a built-in model monitoring	No	No	No	Yes
Support for a fully managed service from model development to deployment	No	No	No	Yes
Availability of a built-in visualizer for analyzing the processed data	No	No	No	Yes
Availability of a predefined environment for ETL logic development	Yes	No	Yes	Yes

Table 5.1 – A comparison of the various data processing setups – a single-node EC2 instance, Glue, EMR, and SageMaker

The right setup depends on both technical and non-technical factors, including the source of the data, the amount of data, the availability of MLOps, and the cost.

> **Things to remember**
>
> a. The four ETL setups we described in this chapter have distinct advantages.
>
> b. When selecting a particular setup, various factors must be considered: the source of the data, the amount of data, the availability of MLOps, and the cost.

Summary

One of the difficulties with DL projects arises from the amount of data. Since a large amount of data is necessary to train a DL model, data processing steps can take up a lot of resources. Therefore, in this chapter, we learned how to utilize the most popular cloud service, AWS, to process terabytes and petabytes of data efficiently. The system includes a scheduler, data storage, databases, visualization, as well as a data processing tool for running the ETL logic.

We have spent extra time looking at ETL since it plays a major role in data processing. We introduced Spark, which is the most popular tool for ETL, and described four different ways of setting up ETL jobs using AWS. The four settings include using a single-node EC2 instance, an EMR cluster, Glue, and SageMaker. Each setup has distinct advantages, and the right one may differ based on the situation. This is because you need to consider both technical and non-technical aspects of the project.

Similar to how the amount of data becomes an issue for processing data, it also introduces multiple issues when training a model. In the next chapter, you will learn how to train models efficiently using a distributed system.

Efficient Model Training

Similar to how we scaled up data processing pipelines in the previous chapter, we can reduce the time it takes to train **deep learning** (DL) models by allocating more computational resources. In this chapter, we will learn how to configure the **TensorFlow** (**TF**) and **PyTorch** training logic to utilize multiple CPU and GPU devices on different machines. First, we will learn how TF and PyTorch support distributed training without any external tools. Next, we will describe how to utilize SageMaker, since it is built to handle the DL pipeline on the cloud from end to end. Lastly, we will look at tools that have been developed specifically for distributed training: Horovod, Ray, and Kubeflow.

In this chapter, we're going to cover the following main topics:

- Training a model on a single machine
- Training a model on a cluster
- Training a model using SageMaker
- Training a model using Horovod
- Training a model using Ray
- Training a model using Kubeflow

Technical requirements

You can download the supplemental material for this chapter from this book's GitHub repository: `https://github.com/PacktPublishing/Production-Ready-Applied-Deep-Learning/tree/main/Chapter_6`.

Training a model on a single machine

As described in *Chapter 3*, *Developing a Powerful Deep Learning Model*, training a DL model involves extracting meaningful patterns from a dataset. When the size of the dataset is small and the model has few parameters to tune, a **central processing unit** (**CPU**) might be sufficient to train the model. However, DL models have shown greater performance when they are trained with a larger training set and consist of a greater number of neurons. Therefore, training using a **graphics processing unit** (**GPU**) has become the standard since you can exploit its massive parallelism in matrix multiplication.

Utilizing multiple devices for training in TensorFlow

TF provides the `tf.distribute.Strategy` module, which allows you to use multiple GPU or CPU devices for training with very simple code modifications (`https://www.tensorflow.org/guide/distributed_training`). `tf.distribute.Strategy` is fully compatible with `tf.keras.Model.fit`, as well as custom training loops, as described in the *Implementing and training a model in TensorFlow* section of *Chapter 3, Developing a Powerful Deep Learning Model*. Various components of Keras, including variables, layers, models, optimizers, metrics, summaries, and checkpoints, are designed to support various `tf.distribute.Strategy` classes, keeping the transition to distributed training as simple as possible. Let's have a look at how the `tf.distribute.Strategy` module allows you to quickly modify a set of code designed for a single device to multiple devices on a single machine:

```
import tensorflow as tf
mirrored_strategy = tf.distribute.MirroredStrategy()
# or
# mirrored_strategy = tf.distribute.
MirroredStrategy(devices=["/gpu:0", "/gpu:1", "/gpu:3"])
# if you want to use only specific devices
with mirrored_strategy.scope():
    # define your model
    # …
model.compile(... )
model.fit(... )
```

Once the model has been saved, it can be loaded with or without the `tf.distribute.Strategy` scope. To achieve distributed training with a custom training loop, you can follow the example presented at `https://www.tensorflow.org/tutorials/distribute/custom_training`. Having said that, let's review the most used strategies. We will cover the most common approaches, some of which go beyond training a single instance. They will be used in the next few sections, which cover training on multiple machines:

- Strategies that provide full support for `tf.keras.Model.fit` and custom training loops:

  - `MirroredStrategy`: Synchronous distributed training using multiple GPUs on a single machine

- `MultiWorkerMirroredStrategy`: Synchronous distributed training on multiple machines (potentially using multiple GPUs per machine). This strategy class requires a TF cluster that's been configured using the `TF_CONFIG` environment variable (https://www.tensorflow.org/guide/distributed_training#TF_CONFIG)

  - `TPUStrategy`: Training on multiple **tensor processing units** (**TPUs**)

- Strategies with experimental features (meaning that classes and methods are still in the development stage) for `tf.keras.Model.fit` and custom training loops:

  - `ParameterServerStrategy`: Model parameters are shared across multiple workers (the cluster consists of workers and parameter servers). Workers read and update the variables that are created on parameter servers after each iteration.

  - `CentralStorageStrategy`: Variables are stored in central storage and replicated across each GPU.

- The last strategy that we want to mention is `tf.distribute.OneDeviceStrategy` (https://www.tensorflow.org/api_docs/python/tf/distribute/OneDeviceStrategy). It runs the training code on a single GPU device as follows:

  ```
  strategy = tf.distribute.OneDeviceStrategy(device="/
  gpu:0")
  ```

  In the preceding example, we have selected the first GPU (`"/gpu:0"`).

It is also worth mentioning that the `tf.distribute.get_strategy` function can be used to get the current `tf.distribute.Strategy` object. You can use this function to change the `tf.distribute.Strategy` object dynamically for your training code, as shown in the following code snippet:

```
if tf.config.list_physical_devices('GPU'):
    strategy = tf.distribute.MirroredStrategy()
else:  # Use the Default Strategy
    strategy = tf.distribute.get_strategy()
```

In the preceding code, we are using `tf.distribute.MirroredStrategy` when GPU devices are available and fall back to the default strategy when GPU devices are not available. Next, let's look at the features provided by PyTorch.

Utilizing multiple devices for training in PyTorch

To train a PyTorch model successfully, the model and input tensor need to be configured for the same device. If you want to use a GPU device, they need to be loaded on the target GPU device explicitly before training, using either the to(device=torch.device('cuda')) or cuda() function:

```
cpu = torch.device(cpu')
cuda = torch.device('cuda')      # Default CUDA device
cuda0 = torch.device('cuda:0')
x = torch.tensor([1., 2.], device=cuda0)
# x.device is device(type='cuda', index=0)
y = torch.tensor([1., 2.]).cuda()
# y.device is device(type='cuda', index=0)
# transfers a tensor from CPU to GPU 1
a = torch.tensor([1., 2.]).cuda()
# a.device are device(type='cuda', index=1)
# to function of a Tensor instance can be used to move the
tensor to different devices
b = torch.tensor([1., 2.]).to(device=cuda)
# b.device are device(type='cuda', index=1)
```

The preceding example shows some of the key operations you should be aware of when using a GPU device. This is a subset of what is presented in the official PyTorch documentation: https://pytorch.org/docs/stable/notes/cuda.html.

However, setting up individual components for training can be tiresome. Therefore, **PyTorch Lightning** (**PL**) has decided to manage this automatically behind the scenes. In the case of PL, target devices can be chosen at the time of training, through the gpus parameter of Trainer:

```
# Train using CPU
Trainer()
# Specify how many GPUs to use
Trainer(gpus=k)
# Specify which GPUs to use
Trainer(gpus=[0, 1])
# To use all available GPUs put -1 or '-1'
Trainer(gpus=-1)
```

In the preceding example, we are describing various training setups for a single machine: training only using CPU devices, training using a set of GPU devices, and training using all GPU devices.

> **Things to remember**
>
> a. TF and PyTorch provide built-in support for training a model using both CPU and GPU devices.
>
> b. Training can be controlled using the `tf.distribute.Strategy` class in TF. When training a model with a single machine, you can use `MirroredStrategy` or `OneDeviceStrategy`.
>
> c. To train a PyTorch model using GPU devices, the model and relevant tensors need to be loaded on the same GPU device manually. PL hides most of the boilerplate code by handling the placements as part of the `Trainer` class.

In this section, we learned how to utilize multiple devices on a single machine. However, there have been many efforts to utilize a cluster of machines for training as there is a limit on the computational power that a single machine can have.

Training a model on a cluster

Even though using multiple GPUs on a single machine has reduced the training time a lot, some models are extremely huge and still require multiple days for training. Adding more GPUs is still an option but physical limitations often exist, preventing you from utilizing the full potential of the multi-GPU setting: motherboards can support a limited number of GPU devices.

Fortunately, many DL frameworks already support training a model on a distributed system. While there are minor differences in the actual implementation, most frameworks adopt the idea of **model parallelism** and **data parallelism**. As shown in the following diagram, model parallelism distributes components of the model to multiple machines, while data parallelism distributes the samples of the training set:

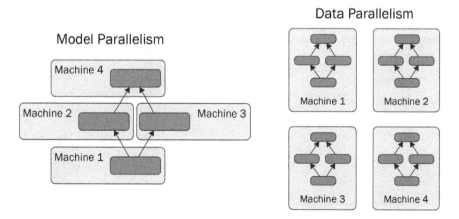

Figure 6.1 – The difference between model parallelism and data parallelism

There are a couple of details that you must be aware of when setting up a distributed system for model training. First, the machines in the cluster need to have a stable connection to the internet since they communicate over the network. If stability is not guaranteed, the cluster must have a way to recover from the connection issue. Ideally, the distributed system should be agnostic to the available machines and be able to add or remove a machine without affecting the overall progress. Such functionality will allow users to increase or decrease the number of machines dynamically, achieving the model training in the most cost-efficient way. AWS provides the aforementioned functionalities out of the box through **Elastic MapReduce (EMR)** and **Elastic Container Service (ECS)**.

In the next two sections, we will take a deeper look into model parallelism and data parallelism.

Model parallelism

In the case of model parallelism, each machine in a distributed system takes a part of the model and manages computations for the assigned components. This approach is often considered when a network is too big to fit on a single GPU. However, it is not that common in reality because GPU devices often have enough memory to fit the model, and it is quite complex to set it up. In this section, we are going to describe the two most basic approaches of model parallelism: **model sharding** and **model pipelining**.

Model sharding

Model sharding is nothing more than partitioning the model into multiple computational subgraphs across multiple devices. Let's assume a simple scenario of a basic single-tier **deep neural network (DNN)** model (no parallel paths). The model can be split into a few consecutive subgraphs, and the sharding profile can be graphically represented as follows. The data will flow sequentially starting from the device with the first subgraph. Each device will pass the computed values to the device of the next subgraph. Until the necessary data arrives, the devices will stay idle. In this example, we have four subgraphs:

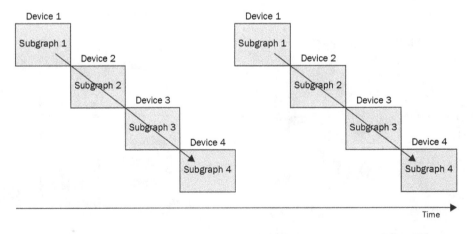

Figure 6.2 – A sample distribution of a model in model sharding; each arrow indicates a mini-batch

As you can see, model sharding does not utilize the full computational resources; a device is waiting for the other device to process its subgraph. To solve this problem, the pipelining approach is proposed.

Model pipelining

In the case of model pipelining, a mini-batch is split into micro-batches and provided to the system in chains, as shown in the following diagram:

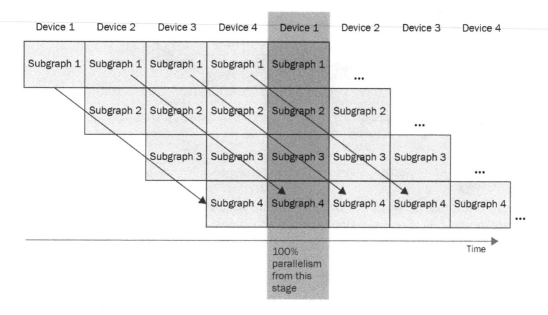

Figure 6.3 – A diagram of model pipeline logic; each arrow indicates a mini-batch

However, model pipelining requires a modified version of backward propagation. Let's look at how a single forward and backward propagation can be achieved in a model pipelining setting. At some point, each device needs to perform not only forward computations for its subgraph but also gradient computations. A single forward and backward propagation can be achieved like so:

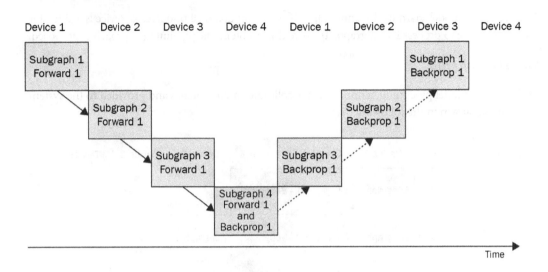

Figure 6.4 – A single forward and backward propagation in model pipelining

In the preceding diagram, we can see that each device runs forward propagation one by one and backward propagation in reverse order, passing the computed values to the next device. Putting everything together, we get the following diagram, which summarizes the logic of model pipelining:

Device 1	Device 2	Device 3	Device 4	Device 1	Device 2	Device 3	Device 4	
Subgraph 1 Forward 1	Subgraph 1 Forward 2	Subgraph 1 Forward 3	Subgraph 1 Forward 4	Subgraph 1 Forward 5	Subgraph 1 Forward 6	Subgraph 1 Backprop 1 Forward 7	Subgraph 1 Backprop 2 Forward 8	...
	Subgraph 2 Forward 1	Subgraph 2 Forward 2	Subgraph 2 Forward 3	Subgraph 2 Forward 4	Subgraph 2 Backprop 1 Forward 5	Subgraph 2 Backprop 2 Forward 6	Subgraph 2 Backprop 3 Forward 7	...
		Subgraph 3 Forward 1	Subgraph 3 Forward 2	Subgraph 3 Backprop 1 Forward 3	Subgraph 3 Backprop 2 Forward 4	Subgraph 3 Backprop 3 Forward 5	Subgraph 3 Backprop 4 Forward 6	...
			Subgraph 4 Forward 1 and Backprop 1	Subgraph 4 Forward 2 and Backprop 2	Subgraph 4 Forward 3 and Backprop 3	Subgraph 4 Forward 4 and Backprop 4	...	

Figure 6.5 – Model parallelism based on model pipelining

To further improve the training time, each device stores the values it computed previously and utilizes them in the following computations.

Model parallelism in TensorFlow

The following code snippet shows how to assign a set of layers to a specific device in TF as you define the model architecture:

```
with tf.device('GPU:0'):
    layer1 = layers.Dense(16, input_dim=8)
with tf.device('GPU:1'):
    layer2 = layers.Dense(4, input_dim=16)
```

If you want to explore model parallelism in TF even more, we recommend checking out the Mesh TF repository (`https://github.com/tensorflow/mesh`).

Model parallelism in PyTorch

Model parallelism is only available on PyTorch and has not yet been implemented in PL. While there are many ways to achieve model parallelism with PyTorch, the most standard approach is to use the `torch.distributed.rpc` module which achieves the communication among the machines using a **remote procedure call** (**RPC**). The three main features of the RPC-based approaches are triggering functions or networks remotely (remote execution), accessing and referencing remote data objects (remote reference), and extending the gradients update functionality of PyTorch across the machine boundaries (distributed gradients update). We delegate the details to the official documentation: `https://pytorch.org/docs/stable/rpc.html`.

Data parallelism

Data parallelism, unlike model parallelism, aims to speed up the training by sharding the dataset to the machines in the cluster. Each machine gets a copy of the model and computes the gradients with the dataset it has been assigned to. Then, the gradients are aggregated and the models are updated globally at once.

Data parallelism in TensorFlow

Data parallelism can be realized in TF by leveraging `tf.distribute.MultiWorkerMirroredStrategy`, `tf.distribute.ParameterServerStrategy`, and `tf.distribute.CentralStorageStrategy`.

We introduced these strategies in the *Utilizing multiple devices for training in TensorFlow* section since specific `tf.distributed` strategies are also used to set up training on multiple devices within a single machine.

To use these strategies, you need to set up a TF cluster where each machine can communicate with the other.

Typically, a TF cluster is defined using a `TF_CONFIG` environment variable. `TF_CONFIG` is just a JSON string that specifies cluster configuration by defining two components: `cluster` and `task`. The following Python code shows how to generate a `.json` file for `TF_CONFIG` from a Python dictionary:

```
tf_config = {
    'cluster': {
        'worker': ['localhost:12345', 'localhost:23456']
    },
    'task': {'type': 'worker', 'index': 0}
}
js_tf = json.dumps(tf_config)
with open("tf_config.json", "w") as outfile:
    outfile.write(js_tf)
```

The `TF_CONFIG` fields and formats are described at `https://cloud.google.com/ai-platform/training/docs/distributed-training-details`.

As demonstrated in the *Utilizing multiple devices for training in TensorFlow* section, you need to put the training code under the `tf.distribute.Strategy` scope. In the following example, we will show a sample usage for `tf.distribute.MultiWorkerMirroredStrategy` class.

First of all, you must put your model instance under the scope of `tf.distribute.MultiWorkerMirroredStrategy`, as shown in the following code snippet:

```
strategy = tf.distribute.MultiWorkerMirroredStrategy()
with strategy.scope():
    model = …
```

Next, you need to make sure the `TF_CONFIG` environment variables have been set up correctly for each machine in the cluster and run the training script, as follows:

```
# On the first node
TF_CONFIG='{"cluster": {"worker": ['localhost:12345',
'localhost:23456']}, "task": {"index": 0, "type": "worker"}}'
python training.py
# On the second node
TF_CONFIG='{"cluster": {"worker": ['localhost:12345',
'localhost:23456']}, "task": {"index": 1, "type": "worker"}}'
python training.py
```

To correctly save your model, please take a look at the official documentation: `https://www.tensorflow.org/tutorials/distribute/multi_worker_with_keras`.

In the case of a custom training loop, you can follow the instructions at `https://www.tensorflow.org/tutorials/distribute/multi_worker_with_ctl`.

Data parallelism in PyTorch

Unlike model parallelism, data parallelism is available for both PyTorch and PL. Among the various implementations, the most standard feature is `torch.nn.parallel.DistributedDataParallel` (DDP). In this section, we will mainly discuss PL as its main advantage comes from the simplicity of the training models that use data parallelism.

To train a model using data parallelism, you need to modify the training code to utilize the underlying distributed system and spawn a process with the `torch.distributed.run` module on each machine (`https://pytorch.org/docs/stable/distributed.html`).

The following code snippet describes what you need to change for ddp. You simply need to provide ddp for the `accelerator` parameter of `Trainer`. `num_nodes` is the parameter to adjust when there is more than one machine in the cluster:

```
# train on 8 GPUs (same machine)
trainer = Trainer(gpus=8, accelerator='ddp')
# train on 32 GPUs (4 nodes)
trainer = Trainer(gpus=8, accelerator='ddp', num_nodes=4)
```

Once the script has been set up, you need to run the following command on each machine. Please keep in mind that `MASTER_ADDR` and `MASTER_PORT` must be consistent as they are used by each processor to communicate. On the other hand, `NODE_RANK` indicates the index of the machine. In other words, it must be different for each machine, and it must start from zero:

```
python -m torch.distributed.run
    --nnodes=2 # number of nodes you'd like to run with
    --master_addr <MASTER_ADDR>
    --master_port <MASTER_PORT>
    --node_rank <NODE_RANK>
    train.py (--arg1 ... train script args...)
```

Based on the official documentation, DDP works as follows:

1. Each GPU across each node spins up a process.
2. Each process gets a subset of the training set.
3. Each process initializes the model.
4. Each process performs both forward and backward propagation in parallel.

5. The gradients are synchronized and averaged across all processes.

6. Each process updates the weights of the model it has.

Things to remember

a. TF and PyTorch provide options for training DL models across multiple machines using model parallelism and data parallelism.

b. Model parallelism splits the model into multiple components and distributes them across machines. To set up model parallelism in TF and PyTorch, you can use the `Mesh TensorFlow` library and the `torch.distributed.rpc` package, respectively.

c. Data parallelism copies the model to each machine and distributes mini-batches across machines for training. In TF, data parallelism can be achieved using either `MultiWorkerMirroredStrategy`, `ParameterServerStrategy`, or `CentralStorageStrategy`. The main package that's been designed for data parallelism in PyTorch is called `torch.nn.parallel.DistributedDataParallel`.

In this section, we learned how to achieve model training where the lifetime of the cluster is explicitly managed. However, some tools manage the clusters for model training as well. Since each of them has different advantages, you should understand the difference to select the right tool for your development.

First, we will look at the built-in features of SageMaker that train a DL model in a distributed fashion.

Training a model using SageMaker

As mentioned in the *Utilizing SageMaker for ETL* section of *Chapter 5, Data Preparation in the Cloud*, the motivation of SageMaker is to help engineers and researchers focus on developing high-quality DL pipelines without worrying about infrastructure management. SageMaker manages data storage and computational resources for you, allowing you to utilize a distributed system for model training with minimal effort. In addition, SageMaker supports streaming data to your models for inferencing, hyperparameter tuning, and tracking experiments and artifacts.

SageMaker Studio is the place where you define the logic for your model. The SageMaker Studio notebooks allow you to quickly explore the available data and set up model training logic. When model training takes too long, scaling up to use multiple computational resources and finding the best set of hyperparameters can be efficiently achieved by making a few modifications to the infrastructure's configuration. Furthermore, SageMaker supports hyperparameter tuning on a distributed system to exploit parallelism.

Even though SageMaker sounds like a magic key for a DL pipeline, there are disadvantages as well. The first is its cost. Instances that have been allocated to SageMaker are around 40% more expensive than equivalent EC2 instances. Next, you may find that not all the libraries are available in the notebook. In other words, you may need to spend some additional time building and installing the library you need.

Setting up model training for SageMaker

By now, you should be able to start a notebook and select a predefined development environment for your project since we covered these in the *Utilizing SageMaker for ETL* section of *Chapter 5, Data Preparation in the Cloud*. Assuming that you have already processed raw data and stored the processed data in a data storage, we will focus on model training in this section. Model training with SageMaker can be summarized into the following three steps:

1. If the processed data in the storage hasn't been split into training, validation, and test sets yet, you must split them first.

2. You need to define the model training logic and specify the cluster configuration.

3. Lastly, you need to train your model and save the artifacts back in data storage. When training is completed, the allocated instances will be terminated automatically.

The key for model training with SageMaker is `sagemaker.estimator.Estimator`. It allows you to configure the training settings, including infrastructure setup, type of Docker images to use, and hyperparameters (`https://sagemaker.readthedocs.io/en/stable/api/training/estimators.html`). The following are the main parameters that you would typically configure:

- `role` (`str`): An AWS IAM role
- `instance_count` (`int`): The number of SageMaker EC2 instances to use for training
- `instance_type` (`str`): The type of SageMaker EC2 instance to use for training
- `volume_size` (`int`): The size of the Amazon **Elastic Block Store** (**EBS**) volume (in gigabytes) that will be used to download input data temporarily for training
- `output_path` (`str`): An S3 object where the training result will be stored
- `use_spot_instances` (`bool`): A flag specifying whether to use SageMaker-managed AWS Spot instances for training
- `checkpoint_s3_uri` (`str`): An S3 URI where the checkpoints will be stored during training
- `hyperparameters` (`dict`): A dictionary containing the initial set of hyperparameters
- `entry_point` (`str`): The path to the Python file to run
- `dependencies` (`list[str]`): A list of directories that will be loaded into the job

So long as you select the right container from Amazon **Elastic Container Registry** (**ECR**), you can set up any training configuration for SageMaker. Containers with various configurations for CPU and GPU devices also exist. You can find these at `https://github.com/aws/deep-learning-containers/blob/master/available_images.md`.

In addition, there exist repositories of open sourced toolkits designed to help TF and PyTorch model training on Amazon SageMaker. These repositories also contain Docker files that already have the necessary libraries installed, such as TF, PyTorch, and other dependencies necessary to build SageMaker images:

- TF: `https://github.com/aws/sagemaker-tensorflow-training-toolkit`
- PyTorch: `https://github.com/aws/sagemaker-pytorch-training-toolkit`

Lastly, we would like to mention that you can build and run the containers on your local machine. You can also update the installed libraries if you need to. If any modification is made, you need to upload the modified container to Amazon ECR before you can use it with `sagemaker.estimator.Estimator`.

In the following two sections, we will describe a set of changes that are required to train TF and PyTorch models.

Training a TensorFlow model using SageMaker

SageMaker provides a `sagemaker.estimator.Estimator` class built for TF: `sagemaker.tensorflow.estimator.TensorFlow` (`https://sagemaker.readthedocs.io/en/stable/frameworks/tensorflow/sagemaker.tensorflow.html`).

The following example shows the wrapper script that you need to write using the `sagemaker.tensorflow.estimator.TensorFlow` class to train a TF model on SageMaker:

```
import sagemaker
from sagemaker.tensorflow import TensorFlow
# Initializes SageMaker session
sagemaker_session = sagemaker.Session()
bucket = 's3://dataset/'
tf_estimator = TensorFlow(entry_point='training_script.py',
                source_dir='.',
                role=sagemaker.get_execution_role(),
                instance_count=1,
                instance_type='ml.c5.18xlarge',
                framework_version=tf_version,
                py_version='py3',
                script_mode=True,
                hyperparameters={'epochs': 30} )
```

Please keep in mind that every key in the hyperparameters parameter must have a corresponding entry defined in ArgumentParser of the training script (train_script.py). In the preceding example, we only have epochs defined ('epochs': 30).

To trigger the training, you need to call the fit function. You will need to provide datasets for training and validation. If you have them on an S3 bucket, the fit function will look as follows:

```
tf_estimator.fit({'training': 's3://bucket/training',
                  'validation': 's3://bucket/validation'})
```

The preceding example will run training_script.py, specified in the entry_point parameter, by locating it in the directory provided by source_dir. The details of the instance can be found in the instance_count and instance_type parameters. The training script will run with the parameters defined for hyperparameters of tf_estimator on the training and validation datasets defined in the fit function.

Training a PyTorch model using SageMaker

Similar to sagemaker.tensorflow.estimator.TensorFlow, there's sagemaker.pytorch.PyTorch (https://sagemaker.readthedocs.io/en/stable/frameworks/pytorch/sagemaker.pytorch.html). You can set up the training for your PyTorch (or PL) model, as described in the *Implementing and training a model in PyTorch* section of *Chapter 5, Data Preparation in the Cloud*, and integrate sagemaker.pytorch.PyTorch, as shown in the following code snippet:

```
import sagemaker
from sagemaker.pytorch import PyTorch
# Initializes SageMaker session
sagemaker_session = sagemaker.Session()
bucket = 's3://dataset/'
pytorch_estimator = PyTorch(
                    entry_point='train.py',
                    source_dir='.',
                    role=sagemaker.get_execution_role(),
                    framework_version='1.10.0',
                    train_instance_count=1,
                    train_instance_type='ml.c5.18xlarge',
                    hyperparameters={'epochs': 6})

...

pytorch_estimator.fit({
```

```
                       'training': bucket+'/training',
                       'validation': bucket+'/validation'})
```

The usage of a PyTorch estimator is identical to a TF estimator described in the previous section.

This concludes the basic usage of SageMaker for model training. Next, we will learn how to scale up training jobs in SageMaker. We will discuss distributed training using a distribution strategy. We will also cover how you can speed up the training by utilizing other data storage services that have lower latency.

Training a model in a distributed fashion using SageMaker

Data parallelism in SageMaker can be achieved using a distributed data parallel library (https://sagemaker.readthedocs.io/en/stable/api/training/smd_data_parallel.html).

All you need to do is to enable dataparallel as you create the sagemaker.estimator. Estimator instance, as follows:

```
distribution = {"smdistributed": {"dataparallel": { "enabled":
True}}
```

The following code snippet shows a TF estimator that's been created with dataparallel. The full details can be found at https://docs.aws.amazon.com/en_jp/sagemaker/latest/dg/data-parallel-use-api.html:

```
tf_estimator = TensorFlow(
                entry_point='training_script.py',
                source_dir='.',
                role=sagemaker.get_execution_role(),
                instance_count=4,
                instance_type='ml.c5.18xlarge',
                framework_version=tf_version,
                py_version='py3',
                script_mode=True,
                hyperparameters={'epochs': 30}
                distributions={'smdistributed':
                "dataparallel": {"enabled": True}})
```

The same modifications are necessary for a PyTorch estimator.

SageMaker supports two different mechanisms for transferring input data to the underlying algorithm: file mode and pipe mode. By default, SageMaker uses file mode, which downloads the input data to an EBS volume for training. However, if the amount of data is huge, this can slow down the training. In this case, you can use pipe mode, which streams data from S3 (using Linux FIFO) without making extra copies.

In the case of TF, you can simply use `PipeModeDataset` from the `sagemaker-tensorflow` extension (https://github.com/aws/sagemaker-tensorflow-extensions) as follows:

```
from sagemaker_tensorflow import PipeModeDataset
ds = PipeModeDataset(channel='training', record_
format='TFRecord')
```

However, training a PyTorch model using pipe mode requires a bit more engineering effort. Therefore, we will point you to a notebook example that describes each step in depth: https://github. com/aws/amazon-sagemaker-examples/blob/main/advanced_functionality/ pipe_bring_your_own/pipe_bring_your_own.ipynb.

The distributed strategy and pipe mode should speed up the training by scaling up the underlying computational resources and increasing the data transfer throughputs. However, if they are not sufficient, you can try leveraging two other more efficient data storage services that are compatible with SageMaker: Amazon **Elastic File System (EFS)** and Amazon **fully managed shared storage (FSx)** which was built for the Lustre filesystem. For more details, you can refer to their official pages at https://aws.amazon.com/efs/ and https://aws.amazon.com/fsx/ lustre/, respectively.

SageMaker with Horovod

The other option for SageMaker distributed training is to use *Horovod,* a free and open source framework for distributed DL training based on **Message Passing Interface** (**MPI**) principles. MPI is a standard message-passing library that is widely used in parallel computing architectures. Horovod assumes that MPI is available for worker discovery and reduction coordination. Horovod can also utilize Gloo instead of MPI, an open source collective communications library. Here is an example of the distribution parameter configured for Horovod:

```
distribution={"mpi": {"enabled":True,
                      "processes_per_host":2 }}
```

In the preceding code snippet, we are achieving coordination among the machines using MPI. `processes_per_host` defines the number of processes to run on each instance. This is equivalent to defining the number of processes using the `-H` parameter in the `mpirun` or `horovodrun` command, which controls the program's execution in MPI and Horovod, respectively.

In the following code snippet, we are selecting the number of parallel processes that control the number of training script executions (the -np parameter). Then, this number is split into specific machines using the specified values for the -H parameter. With the following commands, each machine will run train.py twice. This would be a typical setting when you have four machines with two GPUs each. The sum of assigned -H processes cannot exceed the -np value:

```
mpirun -np 8 -H server1:2,server2:2,server3:2,server4:2 …
(other parameters) python train.py
```

We will discuss Horovod in depth in the following section as we cover how to train a DL model on a standalone Horovod cluster composed of EC2 instances.

> **Things to remember**
>
> a. SageMaker provides an excellent tool, SageMaker Studio, which allows you to quickly perform initial data exploration and train baseline models.
>
> b. The sagemaker.estimator.Estimator object is an important component for training a model using SageMaker. It also supports distributed training on a set of machines with various CPU and GPU configurations.
>
> c. Utilizing SageMaker for TF and PyTorch model training can be achieved estimators that are specifically designed for each framework.

Now, let's look at how to use Horovod without SageMaker for distributed model training.

Training a model using Horovod

Even though we introduced Horovod as we introduced SageMaker, Horovod is designed to support distributed training alone (https://horovod.ai/). It aims to provide a simple way to train models in a distributed fashion by providing nice integrations for popular DL frameworks, including TensorFlow and PyTorch.

As mentioned previously in the *SageMaker with Horovod* section, the core principles of Horovod are based on MPI concepts such as size, rank, local rank, allreduce, allgather, broadcast, and alltoall (https://horovod.readthedocs.io/en/stable/concepts.html).

In this section, we will learn about how to set up a Horovod cluster using EC2 instances. Then, we will describe the modifications you need to make in TF and PyTorch scripts to train your model on the Horovod cluster.

Setting up a Horovod cluster

To set up a Horovod cluster using EC2 instances, you must follow these steps:

1. Go to the EC2 instance console: `https://console.aws.amazon.com/ec2/`.
2. Click the **Launch Instances** button in the top-right corner.
3. Select **Deep Learning AMI** (the abbreviation for Amazon Machine Image) with TF, PyTorch, and Horovod installed. Click the **Next …** button at the bottom right.
4. Select the right **Instance Type** for your training. You can select CPU or GPU instance types that fit your needs. Click the **Next …** button at the bottom right:

1. Choose AMI   **2. Choose Instance Type**   3. Configure Instance   4. Add Storage   5. Add Tags   6. Configu

Step 2: Choose an Instance Type

Amazon EC2 provides a wide selection of instance types optimized to fit different use cases. Instances are virtu
capacity, and give you the flexibility to choose the appropriate mix of resources for your applications. Learn mo

Filter by: p2 ∨   Current generation   ∨   Show/Hide Columns

Currently selected: t2.micro (- ECUs, 1 vCPUs, 2.5 GHz, -, 1 GiB memory, EBS only)

	Family	Type	vCPUs ⓘ	Memory (GiB)
☐	p2	p2.xlarge	4	61
☐	p2	p2.8xlarge	32	488

Figure 6.6 – Instance type selection in the EC2 Instance console

5. Select the desired number of instances that will make up your Horovod cluster. Here, you can also request AWS Spot instances (cheaper instances based on the sparse EC2 capacity that can be interrupted, making them only feasible for fault-tolerant tasks). However, let's use on-demand resources for simplicity.
6. Select the right network and subnet settings. In real life, this type of information will be provided by the DevOps department.
7. On the same page, select **Add instance to placement group** and **Add to a new placement group,** type the name that you want to use for the group, and select **cluster** for **placement group strategy**.
8. On the same page, provide your **Identity and Access Management (IAM)** role so that you can access S3 buckets. Click the **Next …** button at the bottom right.

9. Select the right storage size for your instances. Click the **Next ...** button at the bottom right.

10. Select unique labels/tags (`https://docs.aws.amazon.com/general/latest/gr/aws_tagging.html`) for your instances. In real life, these might be used as additional security measures, such as terminating instances with specific tags. Click the **Next ...** button at the bottom right.

11. Create a security group or choose an existing one. Again, you must talk to the DevOps department to get the proper information. Click the **Next ...** button at the bottom right.

12. Review all the information and launch. You will be asked to provide a **Privacy Enhanced Mail (PEM)** key for authentication.

After these steps, the desired number of instances will start up. If you didn't add the **Name** tag in *Step 10*, your instances will not have any names. In this case, you can navigate to the EC2 Instances console and update the names manually. At the time of writing, you can request static IPv4 addresses called Elastic IPs and assign them to your instances (`https://docs.aws.amazon.com/AWSEC2/latest/UserGuide/elastic-ip-addresses-eip.html`).

Finally, you need to ensure that the instances can communicate with each other without an issue. You should check the **Security Groups** settings and add inbound rules for SSH and other traffic if necessary.

At this point, you just need to copy your PEM key from your local machine to the master EC2 instance. For an Ubuntu AMI, you can run the following command:

```
scp -i <your_pem_key_path> ubuntu@<IPv4_Public_IP>:/home/
ubuntu/.ssh/
```

Now, you can use SSH to connect to the master EC2 instance. What you need to do next is to set the passwordless connections between EC2 instances by providing your PEM key in the SSH command using the following commands:

```
eval 'ssh-agent'
ssh-add <your_pem_key>
```

In the preceding code snippet, the `eval` command sets the environment variables provided by the `ssh-agent` command, while `ssh-add` command adds a PEM identity to the authentication agent.

Now, the cluster is ready to support Horovod! When you are finished, you must stop or terminate your cluster on the web console. Otherwise, it will continuously charge you for the resources.

In the next two sections, we will learn how to change the TF and PyTorch training scripts for Horovod.

Configuring a TensorFlow training script for Horovod

To train a TF model using Horovod, you need the `horovod.tensorflow.keras` module. First of all, you need to import the `tensorflow` and `horovod.tensorflow.keras` modules. We will refer to `horovod.tensorflow.keras` as hvd. Then, you need to initialize the Horovod cluster as follows:

```
import tensorflow as tf
import horovod.tensorflow.keras as hvd
# Initialize Horovod
hvd.init()
```

At this point, you can check the size of the cluster using the `hvd.size` function. Each process in Horovod will be assigned a rank (a number from 0 to the size of the cluster in terms of the processes you want to run or devices you want to use), which you can access through the `hvd.rank` function. On each instance, each process has a distinct number assigned from 0 to the number of processes on that instance, known as the local rank (the unique numbers per instance but duplicated across instances). The local rank for the current process can be accessed using the `hvd.local_rank` function.

You can pin a specific GPU device for each process using local rank as follows. This example also shows how to set memory growth for your GPUs using `tf.config.experimental.set_memory_growth`:

```
gpus = tf.config.experimental.list_physical_devices('GPU')
for gpu in gpus:
    tf.config.experimental.set_memory_growth(gpu, True)
if gpus:
    tf.config.experimental.set_visible_devices(gpus[hvd.local_
rank()], 'GPU')
```

In the following code, we are splitting the data based on rank so that each process trains on a different set of examples:

```
dataset = np.array_split(dataset, hvd.size())[hvd.rank()]
```

For the model architecture, you can follow the instructions in the *Implementing and training a model in TensorFlow* section of *Chapter 3, Developing a Powerful Deep Learning Model*:

```
model = ...
```

Next, you need to configure the optimizer. In the following example, the learning rate is scaled by the Horovod size. Also, the optimizer needs to be wrapped with a Horovod optimizer:

```
opt = tf.optimizers.Adam(0.001 * hvd.size())
opt = hvd.DistributedOptimizer(opt)
```

The next step is to compile your model and put the network architecture definition and optimizer together. When you are calling the `compile` function with a version of TF that's older than v2.2, you need to disable `experimental_run_tf_function` so that TF uses `hvd.DistributedOptimizer` to compute gradients:

```
model.compile(loss=tf.losses.SparseCategoricalCrossentropy(),
              optimizer=opt,
              metrics=['accuracy'],
              experimental_run_tf_function=False)
```

Another component you need to configure is the callback function. You need to add `hvd.callbacks.BroadcastGlobalVariablesCallback(0)`. This will broadcast the initial values of the weights and biases from rank 0 to all other machines and processes. This is necessary to ensure consistent initialization or to correctly restore training from a checkpoint:

```
callbacks=[
    hvd.callbacks.BroadcastGlobalVariablesCallback(0)
]
```

You can use `rank` to perform a particular operation on a specific instance. For example, logging and saving artifacts on a master node can be achieved by checking whether `rank` is 0 (`hvd.rank()==0`), as shown in the following code snippet:

```
# Save checkpoints only on the instance with rank 0 to prevent
other workers from corrupting them.
If hvd.rank()==0:
    callbacks.append(keras.callbacks.ModelCheckpoint('./
checkpoint-{epoch}.h5'))
```

Now, you are ready to trigger the `fit` function. The following example shows how to scale the number of steps per epoch using the size of the Horovod cluster. Messages from the `fit` function will be only visible on the master node:

```
if hvd.rank()==0:
    ver = 1
else:
    ver = 0
model.fit(dataset,
          steps_per_epoch=hvd.size(),
          callbacks=callbacks,
```

```
        epochs=num_epochs,
        verbose=ver)
```

This is all you need to change to train a TF model in a distributed fashion using Horovod. You can find the complete example at https://horovod.readthedocs.io/en/stable/ tensorflow.html. The Keras version can be found at https://horovod.readthedocs. io/en/stable/keras.html. Additionally, you can modify your training script so that it runs in a fault-tolerant way: https://horovod.readthedocs.io/en/stable/elastic_ include.html. With this change, you should be able to use AWS Spot instances and significantly decrease the cost of training.

Configuring a PyTorch training script for Horovod

Unfortunately, PL does not have proper documentation for Horovod support yet. Therefore, we will focus on PyTorch in this section. Similar to what we described in the preceding section, we will demonstrate the code change you need to make for the PyTorch training script. For PyTorch, you need the horovod.torch module, which we will refer to as hvd again. In the following code snippet, we are importing the necessary modules and initializing the cluster:

```
import torch
import horovod.torch as hvd
# Initialize Horovod
hvd.init()
```

As described in the TF example, you need to bind a GPU device for the current process using the local rank:

```
torch.cuda.set_device(hvd.local_rank())
```

The other parts of the training script require similar modifications. The dataset needs to be distributed across the instances using torch.utils.data.distributed.DistributedSampler and the optimizers must be wrapped around hvd.DistributedOptimizer. The major difference comes from hvd.broadcast_parameters(model.state_dict(), root_rank=0), which broadcasts the model weights. You can find the details in the following code snippet:

```
# Define dataset...
train_dataset = ...
# Partition dataset among workers using DistributedSampler
train_sampler = torch.utils.data.distributed.
DistributedSampler(
    train_dataset, num_replicas=hvd.size(), rank=hvd.rank())
```

```
train_loader = torch.utils.data.DataLoader(train_dataset,
batch_size=..., sampler=train_sampler)
# Build model...
model = ...
model.cuda()
optimizer = optim.SGD(model.parameters())
# Add Horovod Distributed Optimizer
optimizer = hvd.DistributedOptimizer(optimizer, named_
parameters=model.named_parameters())
# Broadcast parameters from rank 0 to all other processes.
hvd.broadcast_parameters(model.state_dict(), root_rank=0)
```

Now, you are ready to train the model. The training loop does not require any modifications. You can just pass the input tensor to the model and trigger backward propagation by triggering the `backward` function on the `loss` and `step` function of `optimizer`. The following code snippet describes the main part of the training logic:

```
for epoch in range(num_ephos):
    for batch_idx, (data, target) in enumerate(train_loader):
        optimizer.zero_grad()
        output = model(data)
        loss = F.nll_loss(output, target)
        loss.backward()
        optimizer.step()
```

The complete description can be found on the official Horovod documentation page: `https://horovod.readthedocs.io/en/stable/pytorch.html`.

As the last piece of content for the *Training model using Horovod* section, the next section explains how to use the `horovodrun` and `mpirun` commands to initiate the model training process.

Training a DL model on a Horovod cluster

Horovod uses MPI principles to coordinate work between processes. To run four processes on a single machine, you can use one of the following commands:

```
horovodrun -np 4 -H localhost:4 python train.py
mpirun -np 4 python train.py
```

In both cases, the `-np` parameter defines the number of times the `train.py` script runs in parallel. The `-H` parameter can be used to define the number of processes per machine (see the horovodrun command in the preceding example). As we learn how to run on a single machine, -H can be dropped, as presented in the mpirun command. Other mpirun parameters are described at https://www. open-mpi.org/doc/v4.0/man1/mpirun.1.php#sect6.

If you do not have MPI installed, you can run the horovodrun command using Gloo. To run the same script to localhost four times (four processes) using Gloo, you just need to add the --gloo flag:

```
horovodrun --gloo -np 4 -H localhost:4 python train.py
```

Scaling up to multiple instances is quite simple. The following command shows how to run the training script on four machines using horovodrun:

```
horovodrun -np 4 -H server1:1,server2:1,server3:1,server4:1
python train.py
```

The following command shows how to run the training script on four machines using mpirun:

```
mpirun -np 4 -H server1:1,server2:1,server3:1,server4:1 python
train.py
```

Once one of the preceding commands is triggered from the master node, you will see that each instance runs one process for training.

> **Things to remember**
>
> a. To use Horovod, you need a cluster with open cross-communication among the nodes.
>
> b. Horovod provides a simple and effective way to achieve data parallelism for TF and PyTorch.
>
> c. The training scripts can be executed on a Horovod cluster using the horovodrun or mpirun commands.

In the next section, we will describe Ray, another popular framework for distributed training.

Training a model using Ray

Ray is an open source execution framework for scaling Python workloads across machines (https:// www.ray.io). The following Python workloads are supported by Ray:

- DL model training implemented with PyTorch or TF

- Hyperparameter tuning via Ray Tune (https://docs.ray.io/en/latest/tune/index.html)

- **Reinforcement learning** (**RL**) via RLlib (https://docs.ray.io/en/latest/rllib/index.html), an open source library for RL

- Data processing leveraging Ray Datasets (`https://docs.ray.io/en/latest/data/dataset.html`)

- Model serving via Ray Serve (`https://docs.ray.io/en/latest/serve/index.html`)

- A general Python application leveraging Ray Core (`https://docs.ray.io/en/latest/ray-core/walkthrough.html`)

The key advantage of Ray comes from the simplicity of its cluster definition; you can define a cluster with machines of different types and from various sources. For example, Ray allows you to build instance fleets (clusters based on a wide variety of EC2 instances with flexible and elastic resourcing strategies for each node) by mixing AWS EC2 on-demand instances and EC2 Spot instances with different CPU and GPU configurations. Ray simplifies both cluster creation and integration with DL frameworks, making it an effective tool for distributed DL model training processes.

First, we will learn how to set up a Ray cluster.

Setting up a Ray cluster

You can set up a Ray cluster in two ways:

- **Ray Cluster Launcher**: A tool provided by Ray to help build clusters using instances on cloud services, including AWS, GCP, and Azure

- **Manual cluster construction**: All the nodes need to be connected to the Ray cluster manually

A Ray cluster consists of a head node (master node) and worker nodes. The instances that form the cluster should be configured to communicate with each other over the network. Communication among Ray instances is based on a **Transmission Control Protocol** (**TCP**) connection, and you must have the corresponding ports open. In the next two sections, we will take a closer look at Ray Cluster Launcher and manual cluster construction.

Setting up a Ray cluster using Ray Cluster Launcher

A YAML file is used to configure the cluster when using Ray Cluster Launcher. You can find many sample YAML files for different configurations on Ray's GitHub repository: `https://github.com/ray-project/ray/tree/master/python/ray/autoscaler`.

We will introduce the most basic one in this section. The YAML file starts with some basic information about the cluster, such as the name of the cluster, number of maximum workers, and upscaling speed, as follows:

```
cluster_name: BookDL
max_workers: 5
upscaling_speed: 1.0
```

Next, it configures the cloud service providers:

```
provider:
    type: aws
    region: us-east-1
    availability_zone: us-east-1c, us-east-1b, us-east-1a
    cache_stopped_nodes: True
    ssh_user: ubuntu
    ssh_private_key: /Users/BookDL/.ssh/BookDL.pem
```

In the preceding example, we specify the provider type (type: aws) and select the Region and Availability Zone where instances will be provided (region: us-east-1 and availability_zone: us-east-1c, us-east-1b, us-east-1a). Then, we define whether nodes can be reused in the future (cache_stopped_nodes: True). The last configurations are for user authentication (ssh_user:ubuntu and ssh_private_key:/Users/BookDL/.ssh/BookDL.pem).

Next, the node configuration needs to be specified. First of all, we will start with the head node:

```
available_node_types:
    ray.head.default:
        node_config:
            KeyName:"BookDL.pem"
```

Next, we must set up the security settings. The detailed settings must be consulted with DevOps, which monitors and secures the instances:

```
            SecurityGroupIds:
                - sg-XXXXX
                - sg-XXXXX
            SubnetIds: [subnet-XXXXX]
```

The following configurations are for the instance type and AMI that should be used:

```
            InstanceType: m5.8xlarge
            ImageId: ami-09ac68f361e5f4a13
```

In the following code snippet, we are providing configurations for storage:

```
            BlockDeviceMappings:
                - DeviceName: /dev/sda1
                  Ebs:
                  VolumeSize: 580
```

You can easily define `Tags` as follows:

```
TagSpecifications:
    - ResourceType:"instance"
      Tags:
            - Key:"Developer"
              Value:"BookDL"
```

If needed, you can provide an IAM instance profile for accessing particular S3 buckets:

```
IamInstanceProfile:
    Arn:arn:aws:iam::XXXXX
```

In the next section of the YAML file, we need to provide a configuration for worker nodes:

```
ray.worker.default:
    min_workers: 2
    max_workers: 4
```

First of all, we must specify the number of workers (`min_workers` and `max_workers`). Then, we can define the node configuration similar to how we defined the master node configuration:

```
node_config:
    KeyName: "BookDL.pem"
    SecurityGroupIds:
        - sg-XXXXX
        - sg-XXXXX
    SubnetIds: [subnet-XXXXX]
    InstanceType: p2.8xlarge
    ImageId: ami-09ac68f361e5f4a13
    TagSpecifications:
        - ResourceType: "instance"
          Tags:
                - Key: "Developer"
                  Value: "BookDL"
    IamInstanceProfile:
        Arn: arn:aws:iam::XXXXX
    BlockDeviceMappings:
      - DeviceName: /dev/sda1
```

```
            Ebs:
                VolumeSize: 120
```

In addition, you can specify a list of shell commands to run on each node in the YAML file:

```
setup_commands:
    - (stat $HOME/anaconda3/envs/tensorflow2_p38/ &> /dev/null
&& echo 'export PATH="$HOME/anaconda3/envs/tensorflow2_p38/
bin:$PATH"' >> ~/.bashrc) || true
    - source activate tensorflow2_p38 && pip install --upgrade
pip
    - pip install awscli
    - pip install Cython
    - pip install -U ray
    - pip install -U ray[rllib] ray[tune] ray
    - pip install mlflow
    - pip install dvc
```

In this example, we will add `tensorflow2_p38` for the `conda` environment to the path, activate the environment, and install a few other modules using `pip`. If you want to run some other commands just on the head or worker nodes, you can specify them in `head_setup_commands` and `worker_setup_commands`, respectively. They will be executed after the commands defined in `setup_commands` are executed.

Finally, the YAML file ends with commands for starting the Ray cluster:

```
head_start_ray_commands:
    - ray stop
    - source activate tensorflow2_p38 && ray stop
    - ulimit -n 65536; source activate tensorflow2_p38 &&
ray start --head --port=6379 --object-manager-port=8076
--autoscaling-config=~/ray_bootstrap_config.yaml
worker_start_ray_commands:
    - ray stop
    - source activate tensorflow2_p38 && ray stop
    - ulimit -n 65536; source activate tensorflow2_p38 && ray
start --address=$RAY_HEAD_IP:6379 --object-manager-port=8076
```

At first, setting up a Ray cluster with a YAML file may look complex. However, once you are used to it, you will notice that adjusting cluster settings for future projects becomes rather simple. In addition, it reduces the time needed to spin up correctly defined clusters significantly as you may reuse information about security groups, subnets, tags, and IAM profiles from previous projects.

If you need other details, we recommend you spend some time looking at the official documentation: `https://docs.ray.io/en/latest/cluster/config.html#cluster-config`.

It is worth mentioning that Ray Cluster Launcher supports both autoscaling and using instance fleets with or without EC2 Spot instances. We used AMI in the preceding example, but you can also provide a specific Docker image for your instances. By exploiting the flexibility of the YAML configuration file, you can construct any cluster configurations using a single file.

As we mentioned at the beginning of this section, you can also set up a Ray cluster by manually adding individual instances. We'll look at this option next.

Manually setting up a Ray cluster

Given that you have a set of machines with a network connection, the first step is to install Ray on each machine. Next, you need to change the security settings of each machine so that the machines can communicate with each other. After that, you need to select one node as a head node and run the following command:

```
ray start --head --redis-port=6379
```

The preceding command establishes the Ray cluster; the Redis server (used for the centralized control plane) is started, and its IP address gets printed on the terminal (for example, `123.45.67.89:6379`).

Next, you need to run the following command on all the other nodes:

```
ray start --address=<redis server ip address>
```

The address you need to provide is the one that is printed from the command on the head node.

Now, your machines are ready to support Ray applications. In the manual setting case, the following steps need to be done manually: starting machines, connecting to a head node terminal, copying training files to all nodes, and stopping machines. Let's have a look at how Ray Cluster Launcher can be utilized to help with those tasks.

At this stage, you should be able to specify the desired Ray cluster settings using a YAML file. Whenever you are ready, you can launch your first Ray cluster using the following command:

```
ray up your_cluster_setting_file.yaml
```

To get a remote terminal on the head node, you can run the following command:

```
ray attach your_cluster_setting_file.yaml
```

To terminate the cluster, the following command can be used:

```
ray down your_cluster_setting_file.yaml
```

Now, it's time to learn how to perform DL model training on a Ray cluster.

Training a model in a distributed fashion using Ray

Ray provides the Ray Train library, which allows you to focus on defining training logic by handling the distributed training behind the scenes. Ray Train supports TF and PyTorch. It also provides simple integration with Horovod. In addition, Ray Datasets exists, which provides distributed data loading through distributed data transformations. Finally, Ray provides hyperparameter tuning through the Ray Tune library.

Adjusting TF training logic for Ray is similar to what we described in the *Data parallelism in TensorFlow* section. The main difference comes from the Ray Train library, which helps us set TF_CONFIG.

The adjusted training logic looks as follows:

```
def train_func_distributed():
    per_worker_batch_size = 64
    tf_config = json.loads(os.environ['TF_CONFIG'])
    num_workers = len(tf_config['cluster']['worker'])
    strategy = tf.distribute.MultiWorkerMirroredStrategy()
    global_batch_size = per_worker_batch_size * num_workers
    multi_worker_dataset = dataset(global_batch_size)
    with strategy.scope():
        multi_worker_model = build_and_compile_your_model()
    multi_worker_model.fit(multi_worker_dataset, epochs=20,
steps_per_epoch=50)
```

Then, you can run the training with Ray Trainer, as follows:

```
import ray
from ray.train import Trainer
ray.init()
trainer = Trainer(backend="tensorflow", num_workers=4, use_
gpu=True)
trainer.start()
trainer.run(train_func_distributed)
trainer.shutdown()
```

In the preceding example, the model definition is similar to a single device case, except that it should be compiled with a specific strategy: MultiWorkerMirroredStrategy. The dataset gets split inside the dataset function, providing a different set of samples for each worker node. Finally, the Trainer instance handles the distributed training.

Training PyTorch models using Ray can be achieved with a minimal set of changes as well. A few examples are presented at `https://docs.ray.io/en/latest/train/examples.html#pytorch`.

In addition, you can use Ray with Horovod, where you can leverage Elastic Horovod to train in a fault-tolerant way. Ray will autoscale the training process by simplifying the discovery and orchestration of hosts. We will not cover the details, but a good starting point can be found at `https://docs.ray.io/en/latest/train/examples/horovod/horovod_example.html`.

> **Things to remember**
>
> a. The key advantage of Ray comes from its simplicity of cluster definition.
>
> b. A Ray cluster can be created manually by connecting each machine or using a built-in tool called Ray Cluster Launcher.
>
> c. Ray provides a nice support for autoscaling the training process. It simplifies the discovery and orchestration of hosts.

Finally, let's learn how to use Kubeflow for distributed training.

Training a model using Kubeflow

Kubeflow (`https://www.kubeflow.org`) covers every step of model development, including data exploration, preprocessing, feature extraction, model training, model serving, inferencing, and versioning. Kubeflow allows you to easily scale from a local development environment to production clusters by leveraging containers and Kubernetes, a management system for containerized applications.

Kubeflow might be your first choice for distributed training if your organization is already using the Kubernetes ecosystem.

Introducing Kubernetes

Kubernetes is an open source orchestration platform that's used to manage containerized workloads and services (`https://kubernetes.io`):

- Kubernetes helps with continuous delivery, integration, and deployment.
- It separates development environments from deployment environments. You can construct a container image and develop the application in parallel.
- The container-based approach ensures the consistency of the environment for development, testing, as well as production. The environment will be consistent on a desktop computer or in the cloud, which minimizes the modifications necessary from one step to the other.

We assume that you have Kubeflow and all of its dependencies installed already, along with a running Kubernetes cluster. The steps we will describe in this section are generic enough that they can be used for any cluster settings – **Minikube** (a local version of Kubernetes), AWS **Elastic Kubernetes Service (EKS)**, or a cluster of many nodes. This is the beauty of containerized workloads and services. The local Minikube installation steps can be found online at https://minikube.sigs.k8s.io/docs/start/. For EKS, we direct you to the AWS user guide: https://docs.aws.amazon.com/eks/latest/userguide/getting-started.html.

Setting up model training for Kubeflow

The first step is to package your training code into a container. This can be achieved with a Docker file. Depending on your starting point, you can use containers from the NVIDIA container image space (TF at https://docs.nvidia.com/deeplearning/frameworks/tensorflow-release-notes/running.html or PyTorch at https://docs.nvidia.com/deeplearning/frameworks/pytorch-release-notes/index.html) or containers directly from DL frameworks (TF at https://hub.docker.com/r/tensorflow/tensorflow or PyTorch at https://hub.docker.com/r/pytorch/pytorch).

Let's have a look at an example TF docker file (kubeflow/tf_example_job):

```
FROM tensorflow/tensorflow:latest-gpu-jupyter
RUN pip install minio -upgrade
RUN pip install -upgrade pip
RUN pip install pandas -upgrade
...
RUN mkdir -p /opt/kubeflow
COPY train.py /opt/kubeflow
ENTRYPOINT ["python", "/opt/kubeflow/train.py"]
```

In the preceding Docker definition, the train.py script is a typical TF training script.

For now, we assume that a single machine will be used for training. In other words, it will be a single container job. Given that you have a Docker file and a training script prepared, you can build your container and push it to the repository using the following commands:

```
docker build -t kubeflow/tf_example_job:1.0
docker push kubeflow/tf_example_job:1.0
```

We will use TFJob, a custom component of Kubeflow that contains a **custom resource descriptor (CRD)** which defines how to use resources during training, and a controller which in our case, enables the TF library. TFJob is represented as a YAML file that describes the container image, the script for training, and execution parameters. Let's have a look at a YAML file, `tf_example_job.yaml`, which contains a Kubeflow model training job running on a single machine:

```
apiVersion: "kubeflow.org/v1"
kind: "TFJob"
metadata:
    name: "tf_example_job"
spec:
    tfReplicaSpecs:
        Worker:
            replicas: 1
        restartPolicy: Never
        template:
            specs:
                containers:
                    - name: tensorflow
                      image: kubeflow/tf_example_job:1.0
```

The API version is defined in the first line. Then, the type of your custom resource is listed, `kind: "TFJob"`. The `metadata` field is used to identify your job by giving it a custom name. The cluster is defined in the `tfReplicaSpecs` field. As shown in the preceding example, the script (`tf_example_job:1.0`) will be executed just once (`replicas: 1`).

To deploy the defined `TFJob` to your cluster, you can use the `kubectl` command, as follows:

```
kubectl apply -f tf_example_job.yaml
```

You can monitor your job with the following command (using the name defined in the metadata):

```
kubectl describe tfjob tf_example_job
```

To perform distributed training, you can use TF code with a specific `tf.distribute.Strategy`, create a new container, and modify `TFJob`. We will have a look at the necessary changes for `TFJob` in the next session.

Training a TensorFlow model in a distributed fashion using Kubeflow

Let's assume that we already have the TF training code from `MultiWorkerMirroredStrategy`. For `TFJob` to support this strategy, you need to adjust `tfReplicaSpecs` in the `spec` field. We can define replicas of the following types through the YAML file:

- **Chief (master)**: Orchestrates computational tasks
- **Worker**: Runs computations

- **Parameter server**: Manages storage for model parameters
- **Evaluator**: Runs evaluations during model training

As the simplest example, we will define a worker as one of those that can act as a chief node. Parameter server and evaluator are not obligatory.

Let's look at the adjusted YAML file, tf_example_job_dist.yaml, for the distributed TF training:

```
apiVersion: "kubeflow.org/v1"
kind: "TFJob"
metadata:
    name: "tf_example_job_dist"
spec:
    cleanPodPolicy: None
    tfReplicaSpecs:
        Worker:
            replicas: 4
            restartPolicy: Never
            template:
                specs:
                    containers:
                        - name: tensorflow
                          image: kubeflow/tf_example_job:1.1
```

The preceding YAML file will run the training job based on MultiWorkerMirroredStrategy on a new container, kubeflow/tf_example_job:1.1. We can deploy TFJob to the cluster with the same command:

```
kubectl apply -f tf_example_job_dist.yaml
```

In the next section, we will learn how to use PyTorch with Ray.

Training a PyTorch model in a distributed fashion using Kubeflow

For PyTorch, we just need to change TFJob to PyTorchJob and provide a PyTorch training script. For the training script itself, please refer to the *Data parallelism in PyTorch* section. The YAML file requires the same set of modifications, as shown in the following code snippet:

```
apiVersion: "kubeflow.org/v1
kind: "PyTorchJob"
metadata:
```

```
    name: "pt_example_job_dist"
spec:
    pytorchReplicaSpecs:
        Master:
            replicas: 1
            restartPolicy: Never
            template:
                specs:
                    containers:
                        - name: pytorch
                          image: kubeflow/pt_example_job:1.0
        Worker:
            replicas: 5
            restartPolicy: OnFailure
            template:
                specs:
                    containers:
                        - name: pytorch
                          image: kubeflow/pt_example_job:1.0
```

In this example, we have one master node and five replicas of worker nodes. The complete details can be found at `https://www.kubeflow.org/docs/components/training/pytorch`.

Things to remember

a. Kubeflow allows you to easily scale from a local development environment to large clusters leveraging containers and Kubernetes.

b. `TFJob` and `PyTorchJob` allow you to run TF and PyTorch training jobs in a distributed fashion using Kubeflow, respectively.

In this section, we described how to utilize Kubeflow for training TF and PyTorch models in a distributed fashion.

Summary

By realizing the benefit of parallelism that comes from multiple devices and machines, we have learned about various ways to train a DL model. First, we learned how to use multiple CPU and GPU devices on a single machine. Then, we covered how to utilize the built-in features of TF and PyTorch to achieve the training in a distributed fashion, where the underlying cluster is managed explicitly. After that, we learned how to use SageMaker for distributed training and scaling up. Finally, the last three sections described frameworks that are designed for distributed training: Horovod, Ray, and Kubeflow.

In the next chapter, we will cover model understanding. We will learn about popular techniques for model understanding that provide some insights into what is happening within the model throughout the training process.

7
Revealing the Secret of Deep Learning Models

So far, we have described how to construct and efficiently train a **deep learning** (**DL**) model. However, model training often involves multiple iterations because only rough guidance on how to configure the training correctly for a given task exists.

In this chapter, we will introduce hyperparameter tuning, the most standard process of finding the right training configuration. As we guide you through the steps of hyperparameter tuning, we will introduce popular search algorithms adopted for the tuning process (grid search, random search, and Bayesian optimization). We will also look into the field of Explainable AI, which is the process of understanding what models do during prediction. We will describe the three most common techniques in this domain: **Permutation Feature Importance** (**PFI**), **SHapley Additive exPlanations** (**SHAP**), **Local Interpretable Model-agnostic Explanations** (**LIME**).

In this chapter, we're going to cover the following main topics:

- Obtaining the best performing model using hyperparameter tuning
- Understanding the behavior of the model with Explainable AI

Technical requirements

You can download the supplemental material for this chapter from this book's GitHub repository at `https://github.com/PacktPublishing/Production-Ready-Applied-Deep-Learning/tree/main/Chapter_7`.

Obtaining the best performing model using hyperparameter tuning

As described in *Chapter 3, Developing a Powerful Deep Learning Model*, obtaining a DL model that extracts the right pattern for the underlying task requires multiple components to be configured appropriately. While building the right model architecture often introduces many difficulties, setting up the proper model training is another challenge that most people struggle with.

In **machine learning (ML)**, *a* **hyperparameter** *refers to any parameter that controls the learning process.* In many cases, data scientists often focus on model-relevant hyperparameters such as the number of a particular type of layer, learning rate, or type of optimizer. However, hyperparameters also include data-relevant configurations such as types of augmentation to apply and a sampling strategy for model training. The iterative process of changing a set of hyperparameters, and understanding performance changes, to find the right set of hyperparameters for the target task is called *hyperparameter tuning*. To be precise, you will have a set of hyperparameters that you want to explore. For each iteration, one or more hyperparameters will be configured differently and a new model will be trained with the adjusted setting. After the iterative process, the hyperparameter configuration used for the best model will be the final output.

In this chapter, we will learn various techniques and tools available for hyperparameter tuning.

Hyperparameter tuning techniques

Techniques for hyperparameter tuning can differ by how the values for the target hyperparameters are selected. Out of the various techniques, we will be focusing on the most common ones: **grid search**, **random search**, and **Bayesian optimization**.

Grid search

The most basic approach is called grid search, where *every possible value is evaluated one by one.* For example, if you want to explore a learning rate from 0 to 1 with an increase of 0.25, then grid search will train the model for every possible learning rate (0.25, 0.5, 0.75, and 1) and select the learning rate that generates the best model.

Random search

On the other hand, *random search generates a random value for the hyperparameter and repeats the training until the maximum number of experiments is reached.* If we convert the example in the previous section for random search, we must define the maximum number of experiments and a boundary for the learning rate. In this example, we will set the maximum number as 5 and the boundary as 0 to 1. Then, random search will select a random value between 0 and 1 and train a model with the selected learning rate. This process will be repeated 5 times and the learning rate that generates the best model will be selected as the output of hyperparameter tuning.

To help your understanding, the following diagram summarizes the difference between grid search and random search:

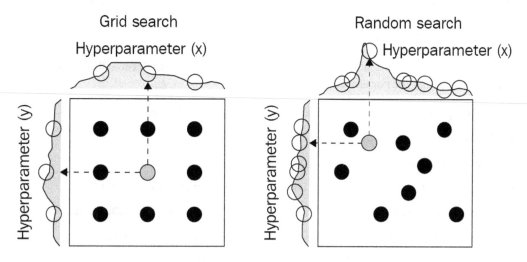

Figure 7.1 – The difference between grid search and random search

In the preceding diagram, x and y indicate two different hyperparameters. The purple and green graphs on each axis represent the model performance changes concerning each hyperparameter.

While grid search and random search are easy to implement, they both have a common limitation: they do not guarantee the best value for the target hyperparameter. This issue mainly comes from the fact that the previous results are not considered when selecting the next value to explore. To overcome this issue, a new search algorithm was introduced: Bayesian optimization.

Bayesian optimization

The idea of Bayesian optimization is straightforward: *a surrogate model that maps the relationship between the hyperparameters and the underlying model is constructed and adjusted throughout the hyperparameter tuning so that we can select a hyperparameter value that will most likely lead us to have a better understanding about the relationship from the following experiment.* Using the generated surrogate model, we can select the hyperparameter value that will likely give us a better model.

There are many ways to build a surrogate model. If we assume that the relationship can be represented as a linear function, the surrogate model generation process will simply be linear regression. In reality, the relationship is much more complex, and the most successful technique is to use Gaussian process regression. Here, we assume that the relationship can be represented by a set of normal distributions. In other words, each value we select is randomly selected from a multivariate normal distribution. We would need to introduce multiple probability and mathematical terms if we wanted to go over every detail of Bayesian optimization. We believe that the high-level description in this section and the complete example in the following section will be sufficient for you to apply hyperparameter tuning using Bayesian optimization. If you would like to understand the theory behind Bayesian optimization, please go to `https://ieeexplore.ieee.org/abstract/document/7352306`.

Hyperparameter tuning tools

As hyperparameter tuning plays an important role in ML projects, many libraries exist that are designed to simplify the process. The popular ones are as follows:

- **Scikit-Optimize**: `https://scikit-optimize.github.io`
- **Optuna**: `https://optuna.org`
- **HyperOpt**: `http://hyperopt.github.io`
- **Ray Tune**: `https://docs.ray.io/en/latest/tune/index.html`
- **Bayesian Optimization**: `https://github.com/fmfn/BayesianOptimization`
- **Metric Optimization Engine (MOE)**: `https://github.com/Yelp/MOE`
- **Spearmint**: `https://github.com/HIPS/Spearmint`
- **GPyOpt**: `https://github.com/SheffieldML/GPyOpt`
- **SigOpt**: `https://sigopt.com`
- **FLAML**: `https://github.com/microsoft/FLAML`
- **Dragonfly**: `https://github.com/dragonfly/dragonfly`
- **HpBandSter**: `https://github.com/automl/HpBandSter`
- **Nevergrad**: `https://github.com/facebookresearch/nevergrad`
- **ZOOpt**: `https://github.com/polixir/ZOOpt`
- **HEBO**: `https://github.com/huawei-noah/HEBO/tree/master/HEBO`
- **SageMaker**: `https://docs.aws.amazon.com/sagemaker/latest/dg/automatic-model-tuning-how-it-works.html`

Among the various tools, we will look at Ray Tune as we covered how to use Ray for distributed training in *Chapter 6, Efficient Model Training*, in the *Training a model using Ray* section.

Hyperparameter tuning using Ray Tune

As part of Ray, a framework developed for scaling Python workloads across machines, Ray Tune is designed for experiment execution and hyperparameter tuning at scale. In this section, we will walk you through how to configure and schedule hyperparameter tuning using **Ray Tune**. Even though the examples are designed for an abstract representation of model training functionality, setups and documentation of Ray Tune are clear enough that **PyTorch** and **TensorFlow** (TF) integration will come naturally at the end of the section.

First, we will look at the basics of Ray Tune. The core functionality of Ray Tune comes from the `tune.run` function, which manages all the experiments, logs, and checkpoints. The basic usage of the `tune.run` function is demonstrated in the following code snippet:

```
from ray import tune
def tr_function(conf):
    num_iterations = conf["num_it"]
    for i in range(num_iterations):
        ... // training logic
        tune.report(mean_accuracy=acc)
tune.run(
    run_or_experiment=tr_function
    conf={"num_it": tune.grid_search([10, 20, 30, 40])})
```

The `tune.run` function takes in `run_or_experiment`, which defines the training logic, and `conf`, which configures the hyperparameter tuning. The number of experiments depends on the type of search function provided for each hyperparameter in `conf`. In the preceding example, we have `tune.grid_search([10, 20, 30, 40])`, which will spin up four experiments, each running the function provided for `run_or_experiment` (`tr_function`) with a distinct value of `num_iterations`. Within `tr_function`, we can access the assigned hyperparameter through the `conf` argument. It is worth mentioning that Ray Tune provides a vast number of sampling methods (`https://docs.ray.io/en/latest/tune/api_docs/search_space.html#tune-sample-docs`).

Ray Tune has integrated many open source optimization libraries as part of `tune.suggest` providing various state-of-the-art search algorithms for hyperparameter tuning. Popular ones include HyperOpt, Bayesian Optimization, Scitkit-Optimize, and Optuna. The complete list can be found at `https://docs.ray.io/en/latest/tune/api_docs/suggestion.html`. In the following example, we will describe how to use `BayesOptSearch`, which, as its name suggests, implements Bayesian optimization:

```
from ray import tune
from ray.tune.suggest.bayesopt import BayesOptSearch
conf = {"num_it": tune.randint(100, 200)}
bayesopt = BayesOptSearch(metric="mean_accuracy", mode="max")
tune.run(
    run_or_experiment=tr_function
```

```
    config = conf,
    search_alg = bayesopt)
```

In the preceding code snippet, we provided an instance of `BayesOptSearch` to the `search_alg` parameter. This example will try to find `num_iterations`, which will provide us a model with the highest `mean_accuracy`.

Another key parameter of `tune.run` is `stop`. This parameter can take in a dictionary, a function, or a `Stopper` object that defines a stopping criterion. If it is a dictionary, keys must be one of the fields in the returned results of the `run_or_experiment` function. If it's a function, it should return a Boolean that becomes `True` once the stopping criteria are met. These two cases are described in the following examples:

```
# dictionary-based stop
tune.run(tr_function,
         stop={"training_iteration": 20,
               "mean_accuracy": 0.96})


# function-based stop
def stp_function(trial_id, result):
    return result["training_iteration"] > 20 or
          result["mean_accuracy"] > 0.96
tune.run(tr_function, stop=stp_function)
```

In the dictionary-based example, each trial will stop if it completes 10 iterations or `mean_accuracy` reaches the specified value, `0.96`. The function-based example implements the same logic but uses the `stp_function` function. For a `stopper` class use case, you can refer to `https://docs.ray.io/en/latest/tune/tutorials/tune-stopping.html#stopping-with-a-class`.

A trial is an internal data structure of Ray Tune that contains metadata about each experiment (`https://docs.ray.io/en/latest/tune/api_docs/internals.html#trial-objects`). Each trial gets a unique ID (`trial.trial_id`) and its hyperparameter settings can be checked through `trial.config`. Interestingly, different scales of machine resources can be allocated for each trial through the `resources_per_trial` parameter of `tune.run` and `trial.placement_group_factory`. Additionally, the `num_samples` parameter can be used to control the number of trials.

The summary of your experiments can be obtained using the `Analysis` instance returned from `ray.tune`. The following code snippet describes a set of information you can retrieve from an `Analysis` instance:

```
# last reported results
df = analysis.results_df
```

```
# list of trials
trs = analysis.trials
# max accuracy
max_acc_df = analysis.dataframe(metric="mean_accuracy",
mode="max")
# dict mapping for all trials in the experiment
all_dfs = analysis.trial_dataframes
```

You can also retrieve other useful information from an Analysis instance. The complete details can be found at https://docs.ray.io/en/latest/tune/api_docs/analysis.html.

This completes the core components of Ray Tune. If you want to integrate Ray Tune for your PyTorch or TF model training, all you must do is adjust tr_function in the examples so that it trains your model as it logs relevant performance metrics.

Overall, we have explored different options for hyperparameter tuning. The tools we have covered in this section should help us efficiently find the best configuration for our DL model.

> **Things to remember**
>
> a. Obtaining a working DL model for a particular task requires finding the right model architecture and using appropriate training configurations. The process of finding the best combination is called hyperparameter tuning.
>
> b. The three most popular hyperparameter tuning techniques are grid search, random search, and Bayesian optimization.
>
> c. Popular hyperparameter tuning tools include Scikit-Optimize, Optuna, Hyperopt, Ray Tune, Bayesian Optimization, MOE, Spearmint, GpyOpt, and SigOpt.

So far, we have treated DL models as black boxes. Hyperparameter tuning involves searching an unknown space that does not explain how the model finds the underlying pattern. In the next section, we will look at what researchers have recently worked on to understand the flexibility of DL.

Understanding the behavior of the model with Explainable AI

Explainable AI is a very active area of research. In business settings, understanding AI models can easily lead to a distinctive competitive advantage. The so-called *black-box models (complex algorithmic models)*, even though they bring exceptional results, are commonly criticized due to their hidden logic. It is hard for higher-level management to fully design the core business based on AI, as interpreting the model and predictions is not an easy task. How can you convince your business partners that an AI model will always deliver the expected results? How can you ensure that the model will still work on new data? How does the model generate the results? Explainable AI helps us address these questions.

Before we go any further, let's look at two important concepts: **interpretability** and **explainability**. At first, they might sound similar. Interpretability tells us why a specific input produces the specific model's output: the effects of specific variables on the result. Explainability goes beyond interpretability; it focuses not only on causality between inputs and outputs but helps us understand how a model works as a whole, including all its sub-elements. Explainability is also driven by three fundamental ideas: transparency, reproducibility, and transferability. This means that we should be able to fully understand what our models do, how data affects the model as it passes through, and be able to reproduce the results.

Explainable AI plays a role in every step of an ML project – development (an explanation of model architecture and meaning of each hyperparameter), training (changes within the model throughout training), as well as inference (results interpretation). In the case of DL models, it is hard to achieve explainability due to the complexity of the network architecture, high algorithmic complexity, and use of random numbers while initializing weights, biases, regularization, and hyperparameters optimization.

In this section, we will discuss a few methods that are commonly used to build additional trustworthiness behind DL models: **Permutation Feature Importance** (PFI), **Feature Importance** (FI), **SHapley Additive exPlanations** (SHAP), and **Local Interpretable Model-agnostic Explanations** (LIME). All these methods are model agnostic; they can be applied to DL models as well as other supporting ML models commonly used to set baseline evaluation metrics.

Permutation Feature Importance

Neural networks lack intrinsic attributes needed to understand the impact of input features on the predictions (the model's output). However, there is a model agnostic approach called **Permutation Feature Importance** (PFI) designed for this difficulty. *The idea of PFI comes from the relationship between input features and outputs: for an input feature that has a high correlation with an output variable, changing its value will increase the model's prediction error.* If the relationship is weak, the model performance won't be affected as much. If the relationship is strong, the performance will be degraded. PFI is often applied to test sets to obtain a broader understanding of the model's interpretability on unseen data.

The key disadvantage of PFI relates to the fact that it will not work correctly when data has a group of correlated input features. In this case, even though you change one feature from the group, the model performance does not change much because other features will remain unchanged.

Going further with the idea, we can completely remove that feature and measure the model performance. This approach is called **Feature Importance** (FI), also known as **Permutation Importance** (PI) or **Mean Decrease Accuracy** (MDA). Let's have a look at how we can implement FI for any black-box model.

Feature Importance

In this section, we will use the *ELI5* Python package (`https://eli5.readthedocs.io`) to perform FI analysis. It stands out in the field of FI because it's very simple to use. Let's look at a minimal code example for TF with a Keras-defined model (see *Chapter 3, Developing a Powerful Deep Learning Model*, for details on model definition):

```
import eli5
from eli5.sklearn import PermutationImportance
def score(self, x, y_true):
    y_pred = model.predict(x)
    return tf.math.sqrt( tf.math.reduce_mean( tf.math.square(y_
pred-y_true), axis=-1))
perm = PermutationImportance(model, random_state=1,
scoring=score).fit(features, labels)
fi_perm=perm.feature_importances_
fi_std=perm.feature_importances_std_
```

As you can see, the code is almost self-explanatory. First, we need to create a wrapper for the score function that calculates the target evaluation metric. Then, the `tf.keras` model gets passed to the constructor of the `PermutationImportance` class. The `fit` function, which takes in features and labels, handles the FI calculation. After this calculation, we can access the mean FI for each feature (`fi_perm`) and the standard deviation of the permuted results (`fi_std`). The following code snippet shows how to visualize the results of permutation importance as a bar graph:

```
plt.figure()
for index, row in enumerate(fi_perm):
    plt.bar(index,
            fi_perm[index],
            color="b",
            yerr=fi_std[index],
            align="center")
plt.show()
```

If the model is neither based on scikit-learn nor Keras, you need to use the `permutation_importance.get_score_importance` function. The following code snippet describes how to use this function with a PyTorch model:

```
import numpy as np
from eli5.permutation_importance import get_score_importances
```

```
# A trained PyTorch model
black_box_model = ...
def score(X, y):
    y_pred = black_box_model.predict(X)
    return accuracy_score(y, y_pred)
base_score, score_decreases = get_score_importances(score, X,
y)
feature_importances = np.mean(score_decreases, axis=0)
```

Unlike the `PermutationImportance` class, the `get_score_importances` function takes in a scoring function, features, and labels all at the same time.

Next, we will have a look at **SHapley Additive exPlanations (SHAP)**, which is also a model-agnostic approach.

SHapley Additive exPlanations (SHAP)

SHAP is an interpretation method that leverages Shapley values to understand the given black-box model. We won't cover the cooperative game theory that SHAP is based on but we will cover the process at a high level. First, let's look at the definition of Shapley values: *the average of marginal contributions among all possible coalitions over different simulations*. What exactly does this mean? Let's say that a group of four friends (*f1, f2, f3*, and *f4*) is working to get the highest score together for an online game. To calculate the Shapley value for a person, we need to calculate the marginal contribution, which is the difference in score when the person is playing versus not playing. This calculation must be done for all possible subgroups (**coalitions**).

Let's take a closer look. To calculate the marginal contribution of *f1* for the coalition of friends *f2, f3*, and *f4*, we need to do the following :

1. Calculate the score (*s1*) generated by all friends (*f1, f2, f3*, and *f4*).
2. Calculate the score (*s2*) generated by friends *f2, f3*, and *f4*.
3. Finally, the marginal contribution of friend *f1* for the coalition of friends *f2, f3*, and *f4 (v)* equals *s1-s2*.

Now, we need to calculate marginal contributions for *all subgroups* (not only for a coalition of friends; that is, *f2, f3*, and *f4*). Here is every possible combination:

1. *f1* versus *no one* is contributing (*v1*)
2. *f1* and *f2* versus *f2* (*v2*)
3. *f1* and *f3* versus *f3* (*v3*)
4. *f1* and *f4* versus *f4* (*v4*)

5. *f1* and *f2* and *f3* versus *f2* and *f3* (*v5*)

6. *f1* and *f2* and *f4* versus *f2* and *f4* (*v6*)

7. *f1* and *f3* and *f4* versus *f3* and *f4* (*v7*)

8. *f1* and *f2* and *f3* and *f4* versus *f2* and *f3* and *f4* (*v8*)

Overall, the Shapley value (*SV*) for *f1* is *(v1+v2+...+v8) / 8*.

For have our results to be statistically sound, we need to run these calculations over multiple simulations. You can see that if we extend the number of friends, the calculations get extremely complex, resulting in high consumption of computational resources. Therefore, specific approximations are used, resulting in different types of so-called explainers (approximators of Shapley values) in the `shap` library (`https://shap.readthedocs.io/en/latest/index.html`). Comparing Shapley's values for all friends, we can find the individual's contribution to the final score.

If we go back to the explanation of DL models, we can see that the friends become a set of features and that the score is the model performance. With this in mind, let's have a look at SHAP explainers, which can be used for DL models:

- `KernelExplainer`: This is the most popular method and is model agnostic. It's based on **Local Interpretable Model-agnostic Explanations** (**LIME**), which we will discuss in the next section.

- `DeepExplainer`: This method is based on the DeepList approach, which decomposes the output on a specific input (`https://arxiv.org/abs/1704.02685`).

- `GradientExplainer`: This method is based on the extension of integrated gradients (`https://arxiv.org/abs/1703.01365`).

For example, we will present a minimalistic code example where SHAP is applied to a TF model. The complete details can be found in the official documentation at `https://shap-lrjball.readthedocs.io/en/latest/index.html`:

```
import shap
# initialize visualization
shap.initjs()
model = … # tf.keras model or PyTorch model (nn.Module)
explainer = shap.KernelExplainer(model, sampled_data)
shap_values = explainer.shap_values(data, nsamples=300)
shap.force_plot(explainer.expected_value, shap_values, data)
shap.summary_plot(shap_values, sampled_data, feature_
names=names, plot_type="bar")
```

For PyTorch models, you will need to wrap your model in a wrapper to convert the input and output into the correct types (`f=lambda x: model(torch.autograd.Variable(torch.from_numpy(x))).detach().numpy()`). In the proceeding example, we have defined `KernelExplainer`, which takes in a DL model and `sampled_data` as inputs. Next, we calculate the SHAP values (approximations of Shapley values) using the `explainer.shap_values` function. In this example, we are using 300 perturbation samples to estimate the SHAP values for the given prediction. If our `sampled_data` contains 100 examples, we will be performing 100*300 model evaluations. Similarly, you can use `GradientExplainer` (`shap.GradientExplainer(model, sampled_data)`) or `DeepExplainer` (`shap.DeepExplainer(model, sampled_data)`). The size of `sampled_data` needs to be big enough to represent the distribution correctly. In the last few lines, we visualize the SHAP values in an additive force layout using the `shap.force_plot` function and create a global model interpretation plot using the `shap.summary_plot` function.

Now, let's look at the LIME approach.

Local Interpretable Model-agnostic Explanations (LIME)

LIME is a method that trains a local surrogate model to explain the model predictions. First, you need to prepare a model you want to interpret and a sample. LIME uses your model to collect predictions from a set of perturbed data and compare them against the original sample to assign similarity weights (higher if predictions are closer to the prediction on the initial sample). LIME fits an intrinsically interpretable surrogate model on the sampled data using a specific number of features weighted by the similarity weights. Finally, LIME treats the surrogate model interpretation as an interpretation of the black-box model for your selected example. To perform LIME analysis, we can use the `lime` package (`https://lime-ml.readthedocs.io`).

Let's have a look at an example designed for a DL model:

```
from lime.lime_tabular import LimeTabularExplainer as Lime
from matplotlib import pyplot as plt
expl = Lime(features, mode='classification', class_names=[0,
1])
# explain first sample
exp = expl.explain_instance(x[0], model.predict, num_
features=5, top_labels=1)
# show plot
exp.show_in_notebook(show_table=True, show_all=False)
```

In the preceding example, we are using the `LimeTabularExplainer` class. The constructor takes in a train set, feature, class names, and a mode type (`'classification'`). Similarly, you can set LIME for regression problems by providing the `'regression'` mode type. Then, by showing the five most important features and their influences, we explain the first prediction from the test set (`x[0]`). Lastly, we generate a plot from the computed LIME explanation.

> **Things to remember**
>
> a. Model interpretability and explainability are the two key concepts in Explainable AI.
>
> b. Popular model-agnostic techniques in Explainable AI are PFI, FI, SHAP, and LIME.
>
> c. PFI, FI, and SHAP are methods that allow you to interpret your model at both local (a single sample) and global (a set of samples) levels. On the other hand, LIME focuses on a single sample and the corresponding model prediction.

In this section, we have explained the idea of Explainable AI and the four most common techniques: PFI, FI, SHAP, and LIME.

Summary

We started the chapter with hyperparameter tuning. We described the three basic search algorithms that are used for hyperparameter tuning (grid search, random search, and Bayesian optimization) and introduced many tools you can integrate into your project. Out of the tools we listed, we covered Ray Tune as it supports distributed hyperparameter tuning and implements many of the state-of-the-art search algorithms out of the box.

Then, we discussed Explainable AI. We explained the most standard techniques (PFI, FI, SHAP, and LIME) and how they can be used to find out how a model's behavior changes with respect to each feature in a dataset.

In the next chapter, we will shift our focus toward deployment. We will learn about ONNX, an open format for ML models, and look at how to convert a TF or PyTorch model into an ONNX model.

Part 3 – Deployment and Maintenance

Many complex challenges often arise during deployment of a deep learning project. In many cases, the deployment settings are different from the development settings and the discrepancy can introduce various restrictions. In this part, we introduce common issues that engineers often struggle with and share effective solutions for each challenge. In the final chapter, we describe the last phase of a deep learning project, which consists of evaluating the project and discussing potential improvements for future projects.

This part comprises the following chapters:

- *Chapter 8, Simplifying Deep Learning Model Deployment*
- *Chapter 9, Scaling a Deep Learning Pipeline*
- *Chapter 10, Improving Inference Efficiency*
- *Chapter 11, Deep Learning on Mobile Devices*
- *Chapter 12, Monitoring Deep Learning Endpoints in Production*
- *Chapter 13, Reviewing the Completed Deep Learning Project*

Simplifying Deep Learning Model Deployment

The **deep learning** (**DL**) models that are deployed in production environments are often different from the models that are fresh out of the training process. They are usually augmented to handle incoming requests with the highest performance. However, the target environments are often too broad, so a lot of customization is necessary to cover vastly different deployment settings. To overcome this difficulty, you can make use of **open neural network exchange** (**ONNX**), a standard file format for ML models. In this chapter, we will introduce how you can utilize ONNX to convert DL models between DL frameworks and how it separates the model development process from deployment.

In this chapter, we're going to cover the following main topics:

- Introduction to ONNX
- Conversion between TensorFlow and ONNX
- Conversion between PyTorch and ONNX

Technical requirements

You can download the supplemental material for this chapter from the following GitHub link: `https://github.com/PacktPublishing/Production-Ready-Applied-Deep-Learning/tree/main/Chapter_8`.

Introduction to ONNX

There are a variety of DL frameworks you can use to train a DL model. However, *one of the major difficulties in DL model deployment arises from the lack of interoperability among these frameworks.* For example, conversion between PyTorch and **TensorFlow** (**TF**) introduces many difficulties.

In many cases, DL models are augmented further for the deployment environment to increase accuracy and reduce inference latency, utilizing the acceleration provided by the underlying hardware. Unfortunately, this requires a broad knowledge of software as well as hardware because each type of hardware provides different accelerations for the running application. Hardware that is commonly used for DL includes the **Central Processing Unit (CPU)**, **Graphical Processing Unit (GPU)**, **Associative Processing Unit (APU)**, **Tensor Processing Unit (TPU)**, **Field Programmable Gate Array (FPGA)**, **Vision Processing Unit (VPU)**, **Neural Processing Unit (NPU)**, and **JetsonBoard**.

This process is not a one-time operation; once the model has been updated in any way, this process may need to be repeated. To reduce the engineering effort in this domain, a group of engineers have worked together to come up with a mediator that standardizes the model components: **ONNX**. This innovative idea helps us train various DL models using any tool without worrying about the difficulties in deployment. Currently, ONNX is the standard file format for **machine learning (ML)** models that enables you to export a fully trained ML model from one framework for other development environments. ONNX generates an `.onnx` file that keeps track of how the model is designed and how each operation within a network is linked to other components. **Netron** is a popular tool that people use to visualize the DL network inside an `.onnx` file (`https://github.com/lutzroeder/netron`). The following is a sample visualization:

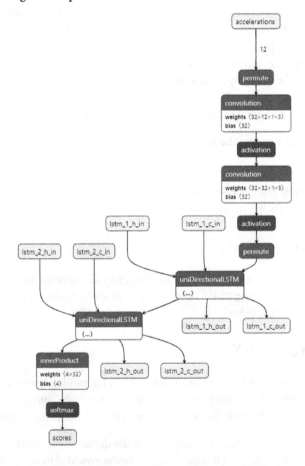

Figure 8.1 – Netron visualization for an ONNX file

As you can see, ONNX is a layer between training frameworks and deployment environments. While the ONNX file defines an exchange format, there also exists **ONNX Runtime (ORT)**, which supports hardware-agnostic acceleration for ONNX models. In other words, the ONNX ecosystem allows you to choose any DL framework for training and makes hardware-specific optimization for deployment easily achievable:

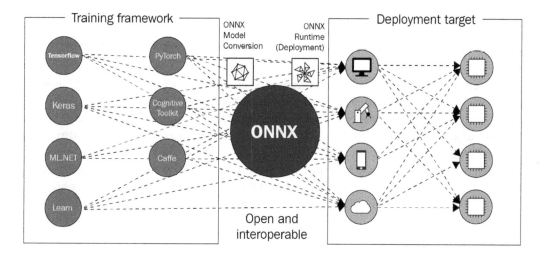

Figure 8.2 – The position of ONNX in a DL project

To summarize, ONNX helps with the following tasks:

- Simplifying the model conversion among various DL frameworks

- Providing hardware-agnostic optimizations for DL models

In the following section, we will take a closer look at ORT.

Running inference using ONNX Runtime

ORT is designed to support training and inferencing using ONNX models directly without converting them into a particular framework. However, training is not the main use case of ORT, so we will focus on the latter aspect, inferencing, in this section.

ORT leverages different hardware acceleration libraries, so-called **Execution Providers (EPs)**, to improve the latency and accuracy of various hardware architectures. The ORT inference code will stay the same regardless of the DL framework used during model training and the underlying hardware.

The following code snippet is a sample ONNX inference code. The complete details can be found at `https://onnxruntime.ai/docs/get-started/with-python.html`:

```
import onnxruntime as rt
providers = ['CPUExecutionProvider'] # select desired provider
or use rt.get_available_providers()
model = rt.InferenceSession("model.onnx", providers=providers)
onnx_pred = model.run(output_names, {"input": x}) # x is your
model's input
```

The `InferenceSession` class takes in a filename, a serialized ONNX model, or an ORT model in a byte string. In the preceding example, we specified the name of an ONNX file (`"model.onnx"`). The `providers` parameter and a list of execution providers ordered by precedence (such as `CPUExecutionProvider`, `TvmExecutionProvider`, `CUDAExecutionProvider`, and many more) are optional but important as they define the type of hardware acceleration that will be applied. In the last line, the `run` function triggers the model prediction. There are two main parameters for the `run` function: `output_names` (the names of the model's output) and `input_feed` (the input dictionary with input names and values that you want to run model prediction with).

Things to remember

a. ONNX provides a standardized and cross-platform representation for ML models.

b. ONNX can be used to convert a DL model implemented in one DL framework into another with minimal effort.

c. ORT provides hardware-agnostic acceleration for deployed models.

In the next two sections, we will look at the process of creating ONNX models using TF and PyTorch.

Conversion between TensorFlow and ONNX

First, we will look at the conversion between TF and ONNX. We will break down the process into two: converting a TF model into an ONNX model and converting an ONNX model back into a TF model.

Converting a TensorFlow model into an ONNX model

`tf2onnx` is used to convert a TF model into an ONNX model (`https://github.com/onnx/tensorflow-onnx`). This library supports both versions of TF (version 1 as well as version 2). Furthermore, conversions to deployment-specific TF formats such as TensorFlow.js and TensorFlow Lite are also available.

To convert a TF model generated using the `saved_model` module into an ONNX model, you can use the `tf2onnx.convert` module, as follows:

```
python -m tf2onnx.convert --saved-model tensorflow_model_path
--opset 9 --output model.onnx
```

In the preceding command, `tensorflow-model-path` points to a TF model saved on disk, `--output` defines where the generated ONNX model will be saved, and `--opset` sets ONNX to opset, which defines the ONNX version and operators (`https://github.com/onnx/onnx/releases`). If your TF model wasn't saved using the `tf.saved_model.save` function, you need to specify the input and output format as follows:

```
# model in checkpoint format
python -m tf2onnx.convert --checkpoint tensorflow-model-
meta-file-path --output model.onnx --inputs input0:0,input1:0
--outputs output0:0
# model in graphdef format
python -m tf2onnx.convert --graphdef tensorflow_model_graphdef-
file --output model.onnx --inputs input0:0,input1:0 --outputs
output0:0
```

The preceding commands describe the conversion for models in Checkpoint (`https://www.tensorflow.org/api_docs/python/tf/train/Checkpoint`) and GraphDef (`https://www.tensorflow.org/api_docs/python/tf/compat/v1/GraphDef`) formats. The key arguments are `--checkpoint` and `--graphdef`, which indicate the model format as well as the location of the source model.

`tf2onnx` also provides a Python API that you can find at `https://github.com/onnx/tensorflow-onnx`.

Next, we will look at how to convert an ONNX model into a TF model.

Converting an ONNX model into a TensorFlow model

While `tf2onnx` is used for conversion from TF into ONNX, `onnx-tensorflow` (`https://github.com/onnx/onnx-tensorflow`) is used for converting an ONNX model into a TF model. It is based on terminal commands as in the case of `tf2onnx`. The following line shows a simple `onnx-tf` command use case:

```
onnx-tf convert -i model.onnx -o tensorflow_model_file
```

In the preceding command, the -i parameter is used to specify the source .onnx file, and the -o parameter is used to specify the output location for the new TF model. Other use cases of the onnx-tf command are well-documented at https://github.com/onnx/onnx-tensorflow/blob/main/doc/CLI.md.

In addition, you can perform the same conversion using a Python API:

```
import onnx
from onnx_tf.backend import prepare
onnx_model = onnx.load("model.onnx")
tf_rep = prepare(onnx_model)
tensorflow-model-file-path = path/to/tensorflow-model
tf_rep.export_graph(tensorflow_model_file_path)
```

In the preceding Python code, the ONNX model is loaded using the onnx.load function and then adjusted for conversion using prepare, which was imported from onnx_tf.backend. Finally, the TF model gets exported and saved to the specified location (tensorflow_model_file_path) using the export_graph function.

> **Things to remember**
>
> a. Conversions from TF into ONNX and from ONNX into TF are performed via onnx-tensorflow and tf2onnx, respectively.
>
> b. Both onnx-tensorflow and tf2onnx support command-line interfaces as well as providing a Python API.

Next, we will describe how the conversions from and to ONNX are performed in PyTorch.

Conversion between PyTorch and ONNX

In this section, we will explain how to convert a PyTorch model into an ONNX model and back again. With the conversion between TF and ONNX covered in the previous section, you should be able to convert your model between TF and PyTorch as well by the end of this section.

Converting a PyTorch model into an ONNX model

Interestingly, PyTorch has built-in support for exporting its model as an ONNX model (https://pytorch.org/tutorials/advanced/super_resolution_with_onnxruntime.html). Given a model, all you need is the torch.onnx.export function as shown in the following code snippet:

```
import torch
pytorch_model = ...
# Input to the model
dummy_input = torch.randn(..., requires_grad=True)
onnx_model_path = "model.onnx"
# Export the model
torch.onnx.export(
    pytorch_model,          # model being run
    dummy_input,            # model input (or a tuple for multiple
inputs)
    onnx_model_path         # where to save the model (can be a
file or file-like object) )
```

The first parameter of torch.onnx.export is a PyTorch model that you want to convert. As the second parameter, you must provide a tensor that represents a dummy input. In other words, this tensor must be the size that the model is expecting as an input. The last parameter is the local path for the ONNX model.

After triggering the torch.onnx.export function, you should see an .onnx file generated at the path you provide (onnx_model_path).

Now, let's look at how to load an ONNX model as a PyTorch model.

Converting an ONNX model into a PyTorch model

Unfortunately, PyTorch does not have built-in support for loading an ONNX model. However, there is a popular library for this conversion called onnx2pytorch (https://github.com/ToriML/onnx2pytorch). Given that this library is installed with a pip command, the following code snippet demonstrates the conversion:

```
import onnx
from onnx2pytorch import ConvertModel
```

```
onnx_model = onnx.load("model.onnx")
pytorch_model = ConvertModel(onnx_model)
```

The key class we need from the `onnx2pytorch` module is `ConverModel`. As shown in the preceding code snippet, we pass an ONNX model into this class to generate a PyTorch model.

Things to remember

a. PyTorch has built-in support for exporting a PyTorch model as an ONNX model. This process involves the `torch.onnx.export` function.

b. Importing an ONNX model into a PyTorch environment requires the `onnx2pytorch` library.

In this section, we described the conversion between ONNX and PyTorch. Since we already know how to convert a model between ONNX and TF, the conversion between TF and PyTorch comes naturally.

Summary

In this chapter, we introduced ONNX, a universal representation of ML models. The benefit of ONNX mostly comes from its model deployment, as it handles environment-specific optimization and conversions for us behind the scenes through ORT. Another advantage of ONNX comes from its interoperability; it can be used to convert a DL model generated with a framework for the other frameworks. In this chapter, we covered conversion for TensorFlow and PyTorch specifically, as they are the two most standard DL frameworks.

Taking another step toward efficient DL model deployment, in the next chapter, we will learn how to use **Elastic Kubernetes Service (EKS)** and SageMaker to set up a model inference endpoint.

9

Scaling a Deep Learning Pipeline

Amazon Web Services (**AWS**) opens many possibilities in **deep learning** (**DL**) model deployments. In this chapter, we will introduce the two most popular services designed for deploying a DL model as an inference endpoint: **Elastic Kubernetes Service** (**EKS**) and **SageMaker**.

In the first half, we will describe the EKS-based approach. First, we will discuss how to create inference endpoints for **TensorFlow** (**TF**) and PyTorch models and deploy them using EKS. We will also introduce the **Elastic Inference** (**EI**) accelerator, which can increase the throughput while reducing the cost. EKS clusters have pods that host the inference endpoints as web servers. As the last topic for EKS-based deployment, we will introduce how the pods can be scaled horizontally for the dynamic incoming traffic.

In the second half, we will introduce SageMaker-based deployment. We will discuss how to create inference endpoints for TF, PyTorch, and ONNX models. Additionally, the endpoints will be optimized using **Amazon SageMaker Neo** and EI accelerators. Then, we will set up automatic scaling for the inference endpoints running on SageMaker. Finally, we will wrap up this chapter by describing how to host multiple models in a single SageMaker inference endpoint.

In this chapter, we are going to cover the following main topics:

- Inferencing using Elastic Kubernetes Service
- Inferencing using SageMaker

Technical requirements

You can download the supplemental material for this chapter from this book's GitHub repository at `https://github.com/PacktPublishing/Production-Ready-Applied-Deep-Learning/tree/main/Chapter_9`.

Inferencing using Elastic Kubernetes Service

EKS is designed to provide Kubernetes clusters for application deployment by simplifying the complex cluster management process (https://aws.amazon.com/eks). The detailed steps for creating an EKS cluster can be found at https://docs.aws.amazon.com/eks/latest/userguide/create-cluster.html. In general, an EKS cluster is used to deploy any web service application and scale it as necessary. The inference endpoint on EKS is just a web service application that handles model inference requests. In this section, you will learn how to host a DL model inference endpoint on EKS.

A Kubernetes cluster has a control plane and a set of nodes. The control plane makes scheduling and scaling decisions based on the volume of incoming traffic. With scheduling, the control plane manages which node runs a job at a given point in time. With scaling, the control plane increases or decreases the size of the pod based on the volume of traffic coming into the endpoints. EKS manages these components behind the scenes so that you can focus on hosting your services efficiently and effectively.

This section begins by describing how to set up an EKS cluster. Then, we will describe how to create endpoints using TF and PyTorch to handle model inference requests on an EKS cluster. Next, we will discuss the EI accelerator, which improves the inference performance, along with cost reduction. Finally, we will introduce a way to scale the services dynamically based on the volume of incoming traffic.

Preparing an EKS cluster

The first step of model deployment based on EKS is to create a pod of appropriate hardware resources. In this section, we will use the GPU Docker images recommended by AWS (https://github.com/aws/deep-learning-containers/blob/master/available_images.md). These standard images are already registered and available on **Elastic Container Registry** (**ECR**), which provides a secure, scalable, and reliable registry for Docker images (https://aws.amazon.com/ecr). Next, we should apply the NVIDIA device plugin to the container. This plugin enables **machine learning** (**ML**) operations to exploit the underlying hardware to achieve lower latency. For more details on the NVIDIA device plugin, we recommend reading https://github.com/awslabs/aws-virtual-gpu-device-plugin.

In the following code snippet, we will use kubectl, the **command-line inference** (**CLI**) for Kubernetes, to setup an EKS cluster with NVIDIA device plugin. When managing a Kubernetes cluster through kubectl, you need to provide a YAML file that consists of information about clusters, users, namespaces, and authentication mechanisms (https://kubernetes.io/docs/concepts/configuration/organize-cluster-access-kubeconfig). The most popular operation is kubectl apply, which creates or modifies resources in an EKS cluster:

```
kubectl apply -f https://raw.githubusercontent.com/NVIDIA/
k8s-device-plugin/v1.12/nvidia-device-plugin.yml
```

In the preceding use case, the `kubectl apply` command applies the NVIDIA device plugin according to the specification specified in the YAML file to the Kubernetes cluster.

Configuring EKS

A YAML file is used to configure both the machines that make up the Kubernetes cluster and the application running within the cluster. The configurations in the YAML file can be broken down into two parts based on their type: *deployment* and *service*. The deployment part controls the application running within the pod. In this section, it will be used to create an endpoint from DL models. In the EKS context, a set of applications running on one or more pods of a Kubernetes cluster is called a service. The service part creates and configures the service on the cluster. Throughout the service part, we will create a unique URL for the service that external connections can use and configure load balancing for incoming traffic.

When managing an EKS cluster, namespaces can be useful as they isolate a group of resources within the cluster. To create a namespace, you can simply use the `kubectl create namespace` terminal command, as follows:

```
kubectl create namespace tf-inference
```

In the preceding command, we constructed the `tf-inference` namespace for the inference endpoints and services that we will be creating in the following section.

Creating an inference endpoint using the TensorFlow model on EKS

In this section, we will describe an EKS configuration file (`tf.yaml`) designed to host an inference endpoint using a TF model. The endpoint is created by *TensorFlow Service*, a system designed for deploying a TF model (`https://www.tensorflow.org/tfx/guide/serving`). Since our main focus is on EKS configurations, we will simply assume that a trained TF model is already available on S3 as a `.pb` file.

First, let's look at the `Deployment` part of the configuration, which handles the endpoint creation:

```
kind: Deployment
  apiVersion: apps/v1
  metadata:
    name: tf-inference # name for the endpoint / deployment
    labels:
      app: demo
      role: master
  spec:
```

```
        replicas: 1 # number of pods in the cluster
        selector:
          matchLabels:
            app: demo
            role: master
```

As we can see, the `Deployment` part of the configuration starts with `kind: Deployment`. In this first part of the configuration, we provide some metadata about the endpoint and define the system settings by filling in the `spec` section.

The most important configurations for the endpoint are specified under `template`. We will create an endpoint that can be accessed using **HyperText Transfer Protocol (HTTP)** requests, as well as **Remote Procedure Call (gRPC)** requests. HTTP is the most basic transfer data protocol for web clients and servers. Built on top of HTTP, gRPC is an open source protocol for sending requests and receiving responses in binary format:

```
        template:
          metadata:
            labels:
              app: demo
              role: master
          spec:
            containers:
            - name: demo
              image: 763104351884.dkr.ecr.us-east-1.amazonaws.com/
tensorflow-inference:2.1.0-gpu-py36-cu100-ubuntu18.04 # ECR
image for TensorFlow inference
              command:
              - /usr/bin/tensorflow_model_server # start inference
endpoint
              args: # arguments for the inference serving
              - --port=9000
              - --rest_api_port=8500
              - --model_name=saved_model
              - --model_base_path=s3://mybucket/models
              ports:
              - name: http
                containerPort: 8500 # HTTP port
              - name: gRPC
                containerPort: 9000 # gRPC port
```

Under the `template` section, we specify an ECR image to use (`image: 763104351884.dkr.ecr.us-east-1.amazonaws.com/tensorflow-inference:2.1.0-gpu-py36-cu100-ubuntu18.04`), the command to create the TF inference endpoint (`command: /usr/bin/tensorflow_model_server`), the arguments for TF serving (`args`), and the ports configuration for containers (`ports`).

The TF serving arguments contains the model's name (`--model_name=saved_model`), the location of the model on S3 (`--model_base_path=s3://mybucket/models`), the ports for HTTP access (`--rest_api_port=8500`), and the ports for gRPC access (`--port=9000`). The two `ContainerPort` configurations under `ports` are used to expose the endpoints to external connections (`containerPort: 8500` and `containerPort: 9000`).

Next, let's look at the second part of the YAML file – that is, the configurations for `Service`:

```
kind: Service
  apiVersion: v1
  metadata:
    name: tf-inference # name for the service
    labels:
      app: demo
  spec:
    Ports:
    - name: http-tf-serving
      port: 8500 # HTTP port for the webserver inside the pods
      targetPort: 8500 # HTTP port for access outside the pods
    - name: grpc-tf-serving
      port: 9000 # gRPC port for the webserver inside the pods
      targetPort: 9000 # gRPC port for access outside the pods
    selector:
      app: demo
      role: master
    type: ClusterIP
```

The `Service` part of the configuration starts with `kind: Service`. Under the `name: http-tf-serving` section, we have `port: 8500`, which refers to the port that the TF serving web server is listening to inside the pods for HTTP requests. `targetPort` specifies the port that the pods use to expose the corresponding port. We have another set of ports configuration for gRPC under the `name: grpc-tf-serving` section.

To apply the configuration to the underlying cluster, you can simply provide this YAML file to the `kubectl apply` command.

Next, we will create an endpoint for a PyTorch model on EKS.

Creating an inference endpoint using a PyTorch model on EKS

In this section, you will learn how to create a PyTorch model inference endpoint on EKS. First, we would like to introduce *TorchServe*, an open source model serving framework for PyTorch (https://pytorch.org/serve). It is designed to simplify the process of PyTorch model deployment at scale. EKS configurations for PyTorch model deployment are very similar to what we have described for deploying a TF model in the previous section.

First, a PyTorch model .pth file needs to be converted into a .mar file, which is the format required by TorchServe (https://github.com/pytorch/serve/blob/master/model-archiver/README.md). The conversion can be achieved using the torch-model-archiver package. TorchServe and torch-model-archiver can be downloaded and installed through pip, as follows:

```
pip install torchserve torch-model-archiver
```

The conversion, when using the torch-model-archiver command, is shown in the following code:

```
torch-model-archiver --model-name archived_model --version 1.0
--serialized-file model.pth --handler run_inference
```

In the preceding code, the torch-model-archiver command takes in model-name (the name of the output .mar file, which is archived_model), version (PyTorch version 1.0), serialized-file (the input PyTorch .pth file, which is model.pth), and handler (the name of the file that defines TorchServe inference logic; that is, run_inference, which indicates the file named run_inference.py). The command will generate an archived_model.mar file, which will be uploaded to an S3 bucket for endpoint hosting through EKS.

Another command we would like to introduce before discussing EKS configuration is mxnet-model-server. This command is available in a DLAMI instance, allowing you to host a web server that runs PyTorch inference for the incoming requests:

```
mxnet-model-server --start --mms-config /home/model-server/
config.properties --models archived_model=https://dlc-samples.
s3.amazonaws.com/pytorch/multi-model-server/archived_model.mar
```

In the preceding example, the mxnet-model-server command, with the start parameter, creates an endpoint for the model provided through the models parameter. As you can see, the models parameter points to the location of the model on S3 (archived_model=https://dlc-samples.s3.amazonaws.com/pytorch/multi-model-server/archived_model.mar). The input arguments for the model are specified in the /home/model-server/config.properties file, which is passed to the command through the mms-config parameter.

Now, we will discuss how the `Deployment` part of the EKS configuration must be filled in. Every component can stay similar to the version for the TF model. The main difference comes from the `template` section, as shown in the following code snippet:

```
containers:
- name: pytorch-service
  image: "763104351884.dkr.ecr.us-east-1.amazonaws.com/
pytorch-inference:1.3.1-gpu-py36-cu101-ubuntu16.04"
  args:
  - mxnet-model-server
  - --start
  - --mms-config /home/model-server/config.properties
  - --models archived_model=https://dlc-samples.
s3.amazonaws.com/pytorch/multi-model-server/archived_model.mar
  ports:
  - name: mms
    containerPort: 8080
```

In the preceding code, we are using a different Docker image that has PyTorch installed (`image:"763104351884.dkr.ecr.us-east-1.amazonaws.com/pytorch-inference:1.3.1-gpu-py36-cu101-ubuntu16.04"`). The configuration takes in the `mxnet-model-server` command to create an inference endpoint. The port we will be using for this endpoint is `8080`. The only change we made for the `Service` part can be found in the `Ports` section; we must ensure that an external port is assigned and connected to port `8080` – that is, the port that the endpoint is hosted on. Again, you can use the `kubectl apply` command to apply the changes.

In the next section, we will describe how to interact with the endpoint hosted by the EKS cluster.

Communicating with an endpoint on EKS

Now that we have an endpoint running, we will explain how you can send a request and retrieve an inference result. First, we need to identify the IP address of the service using `kubectl get services`, as shown in the following code snippet:

```
kubectl get services --all-namespaces -o wide
```

The preceding command will return a list of services and their external IP address:

```
NAME          TYPE      CLUSTER-IP    EXTERNAL-IP     PORT(S)   AGE
tf-inference ClusterIP 10.3.xxx.xxx 104.198.xxx.xx 8500/TCP 54s
```

In this example, we will make use of the `tf-inference` service we created in the *Creating an inference endpoint using the TensorFlow model on EKS* section. From the sample output of `kubectl get services`, we can see that the service is running with an external IP address of `104.198.xxx.xx`. To access the service via HTTP, you need to append the port for HTTP to the IP address: `http://104.198.xxx.xx:8500`. If you are interested in creating an explicit URL for the IP address, please go to `https://aws.amazon.com/premiumsupport/knowledge-center/eks-kubernetes-services-cluster`.

To send a prediction request to the endpoint and receive an inference result, you need to make a POST-typed HTTP request. If you want to send a request from the terminal, you can use the `curl` command as follows:

```
curl -d demo_input.json -X POST http://104.198.xxx.xx:8500/v1/
models/demo:predict
```

In the preceding command, we are sending JSON data (`demo_input.json`) to the endpoint (`http://104.198.xxx.xx:8500/v1/models/demo:predict`). The input JSON file, `demo_input.json`, consists of the following code snippet:

```
{
    "instances": [1.0, 2.0, 5.0]
}
```

The response data we will receive from the endpoint also consists of JSON data that looks as follows:

```
{
    "predictions": [2.5, 3.0, 4.5]
}
```

A detailed explanation of the input and output JSON data structures can be found in the official documentation: `https://www.tensorflow.org/tfx/serving/api_rest`.

If you are interested in using gRPC instead of HTTP, you can find the details at `https://aws.amazon.com/blogs/opensource/the-versatility-of-grpc-an-open-source-high-performance-rpc-framework`.

Congratulations! You have successfully created an endpoint for your model that your application can access over the network. Next, we will introduce Amazon EI accelerator, which can reduce the inference latency and EKS costs.

Improving EKS endpoint performance using Amazon Elastic Inference

In this section, we will describe how to create an EKS cluster with the EI accelerator, a low-cost GPU-powered acceleration. The EI accelerator can be linked to Amazon EC2 and Sagemaker instances or **Amazon Elastic Container Service (ECS)** tasks. It reduces the cost of running the DL model by up to 75%. To use the EI accelerator for an EKS cluster, the cluster must be set up with `eia2.*`-typed instances. The complete description of eia2.* instances can be found at `https://aws.amazon.com/machine-learning/elastic-inference/pricing`.

To make the most out of AWS resources, you also need to compile your model using *AWS Neuron* (`https://aws.amazon.com/machine-learning/neuron`). The advantage of Neuron models comes from the fact that they can utilize Amazon EC2 Inf1 instances. These types of machines consist of *AWS Inferentia*, a custom chip designed by AWS for ML in the cloud (`https://aws.amazon.com/machine-learning/inferentia`).

The AWS Neuron SDK is pre-installed in AWS DL containers and **Amazon Machine Images (AMI)**. In this section, we will focus on TF models. However, PyTorch model compilation goes through the same process. The detailed steps for TF can be found at `https://docs.aws.amazon.com/dlami/latest/devguide/tutorial-inferentia-tf-neuron.html` and the steps for PyTorch can be found at `https://docs.aws.amazon.com/dlami/latest/devguide/tutorial-inferentia-pytorch-neuron.html`.

Compiling a TF model into a Neuron model can be achieved by using `tf.neuron.saved_model.compile` function of TF:

```
import tensorflow as tf
tf.neuron.saved_model.compile(
    tf_model_dir, # input TF model dir
    neuron_model_dir # output neuron compiled model dir
)
```

For this function, we simply need to provide where the input model is located (`tf_model_dir`) and where we want to store the output Neuron model (`neuron_model_dir`). Just as we upload a TF model to an S3 bucket for endpoint creation, we need to move the Neuron model to an S3 bucket as well.

Again, the changes you need to make to the EKS configuration only need to be done in the `template` section of the `Deployment` part. The following code snippet describes the updated sections of the configuration:

```
        containers:
        - name: neuron-demo
```

```
      image: 763104351884.dkr.ecr.us-east-1.amazonaws.com/
tensorflow-inference-neuron:1.15.4-neuron-py37-ubuntu18.04
      command:
      - /usr/local/bin/entrypoint.sh
      args:
      - --port=8500
      - --rest_api_port=9000
      - --model_name=neuron_model
      - --model_base_path=s3://mybucket/neuron_model/
      ports:
      - name: http
        containerPort: 8500 # HTTP port
      - name: gRPC
        containerPort: 9000 # gRPC port
```

The first thing we notice from the preceding configuration is that it is very similar to the one we described in the *Creating an inference endpoint using the TensorFlow model on EKS* section. The difference mainly comes from the `image`, `command`, and `args` sections. First, we need to use a DL container with AWS Neuron and TensorFlow Serving applications (`image: 763104351884. dkr.ecr.us-east-1.amazonaws.com/tensorflow-inference-neuron:1.15.4-neuron-py37-ubuntu18.04`). Next, the entry point script for the model artifact file is passed through the `command` key: `/usr/local/bin/entrypoint.sh`. The entry point script is used to start the web server using `args`. To create an endpoint from a Neuron model, we must specify the S3 bucket where the target Neuron model is stored as a `model_base_path` parameter (`--model_base_path=s3://mybucket/neuron_model/`).

To apply the changes to the cluster, you can simply pass the updated YAML file to the `kubectl apply` command.

Lastly, we will look at the autoscaling feature of EKS to increase the stability of the endpoint.

Resizing EKS cluster dynamically using autoscaling

An EKS cluster can automatically adjust the size of the cluster based on the volume of traffic. The idea of horizontal pod autoscaling is to scale up the number of running applications by increasing the number of pods as the number of incoming requests increases. Similarly, some pods will be freed up when the volume of the incoming traffic decreases.

Once an application has been deployed through the `kubectl apply` command, autoscaling can be set up using the `kubectl autoscale` command, as follows:

```
kubectl autoscale deployment <application-name>
--cpu-percent=60 --min=1 --max=10
```

As shown in the preceding example, the `kubectl autoscale` command takes in the name of the application specified in the `Deployment` part of the YAML file, `cpu-percent` (the cut-off CPU percentage that is used to scale up or down the cluster size), `min` (the minimum number of pods to keep), and `max` (the maximum number of pods to spin up). To summarize, the example command will run the service using 1 to 10 pods, depending on the volume of the traffic, keeping the CPU usage at 60%.

Things to remember

a. EKS is designed to provide Kubernetes clusters for application deployment by simplifying the complex cluster management for dynamic traffic.

b. A YAML file is used to configure both the machines that make up the Kubernetes cluster and the application running within the cluster. The two parts of the configuration, `Deployment` and `Service`, control the application running within the pod and configure the service for the underlying target cluster, respectively.

c. It is possible to create and host inference endpoints using TF and PyTorch models on an EKS cluster.

d. By exploiting the EI accelerator with a model compiled using AWS Neuron, it is possible to improve the inference latency while saving the operating cost of the EKS cluster.

b. An EKS cluster can be configured to resize itself dynamically based on the volume of the traffic.

In this section, we discussed EKS-based DL model deployment for TF and PyTorch models. We described how the AWS Neuron model and the EI accelerator can be used to improve service performance. Finally, we covered autoscaling to utilize the available resources more effectively. In the next section, we will look at another AWS service for hosting inference endpoints: SageMaker.

Inferencing using SageMaker

In this section, you will learn how to create an endpoint using SageMaker instead of the EKS cluster. First, we will describe framework-independent ways of creating inference endpoints (the `Model` class). Then, we will look at creating TF endpoints using `TensorFlowModel` and the TF-specific `Estimator` class. The next section will focus on endpoint creation for PyTorch models using the `PyTorchModel` class and the PyTorch-specific `Estimator` class. Furthermore, we will introduce how to build an endpoint from an ONNX model. At this point, we should have a service running model prediction for incoming requests. After that, we will describe how to improve the quality of a service using *AWS SageMaker Neo* and the EI accelerator. Finally, we will cover autoscaling and describe how to host multiple models on a single endpoint.

As described in the *Utilizing SageMaker for ETL* section in *Chapter 5, Data Preparation in the Cloud*, SageMaker provides a built-in notebook environment called SageMaker Studio. The code snippets we have included in this section are meant to be executed in this notebook.

Setting up an inference endpoint using the Model class

In general, SageMaker provides three different classes for endpoint creation. The most basic one is the Model class, which supports models from various DL frameworks. The other option is to use a framework-specific Model class. The last option is to use the Estimator class. In this section, we will look at the first option, which is the Model class.

Before we dive into the endpoint creation process, we need to make sure the necessary components have been prepared appropriately; the right IAM role must be configured for SageMaker, and the trained model should be available on S3. The IAM role can be prepared in the notebook as follows:

```
from sagemaker import get_execution_role
from sagemaker import Session
# IAM role of the notebook
role = get_execution_role()
# A Session object for SageMaker
sess = Session()
# default bucket object
bucket = sess.default_bucket()
```

In the preceding code, the IAM access role and default bucket have been set up. To load the current IAM role of the SageMaker notebook, you can use the sagemaker.get_execution_role function. To create a SageMaker session, you need to create an instance for the Session class. The default_bucket method of the Session instance will create a default bucket with its name in sagemaker-{region}-{aws-account-id} format.

Before uploading the model to an S3 bucket, the model needs to be compressed as a .tar file. The following code snippet describes how to compress the model and upload the compressed model to the target bucket within the notebook:

```
import tarfile
model_archive = "model.tar.gz"
with tarfile.open(model_archive, mode="w:gz") as archive:
    archive.add("export", recursive=True)
# model artifacts uploaded to S3 bucket
model_s3_path = sess.upload_data(path=model_archive, key_
prefix="model")
```

In the preceding code snippet, the compression is performed using the tarfile library. The upload_data method of the Session instance is used to upload the compiled model to the S3 bucket linked with the SageMaker session.

Now, we are ready to create an instance of the Model class. In this particular example, we will assume that the model has been trained with TF:

```python
from sagemaker.tensorflow.serving import Model
# TF version
tf_framework_version = "2.8"
# Model instance for inference endpoint creation
sm_model = Model(
    model_data=model_s3_path, # S3 path for model
    framework_version=tf_framework_version, # TF version
    role=role) # IAM role of the notebook
predictor = sm_model.deploy(
    initial_instance_count=1, # number of instances used
    instance_type="ml.c5.xlarge")
```

As shown in the preceding code, the constructor of the Model class takes in model_data (the S3 path where the compressed model file is located), framework_version (a version of TF), and role (the IAM role for the notebook). The deploy method of the Model instance handles the actual endpoint creation. It takes in initial_instance_count (the number of instances to start the endpoint with) and instance_type (the EC2 instance type to use).

Additionally, you can provide a defined image and drop framework_version. In this case, the endpoint will be created with the Docker image specified for the image parameter. It should be pointing at an image on ECR.

Next, we will discuss how to trigger a model inference from the notebook using the created endpoint. The deploy method will return a Predictor instance. As shown in the following code snippet, you can achieve this through the predict function of the Predictor instance. All you need to pass to this function is some JSON data representing the input:

```python
input = {
    "instances": [1.0, 2.0, 5.0]
}
results = predictor.predict(input)
```

The output of the predict function, results, consists of JSON data that, in our example, looks as follows:

```python
{
    "predictions": [2.5, 3.0, 4.5]
}
```

The predict function supports data of different formats such as JSON, CSV, and multidimensional array. If you need to use a type other than JSON, you can refer to https://sagemaker.readthedocs. io/en/stable/frameworks/tensorflow/using_tf.html#tensorflow-serving-input-and-output.

Another option for triggering model inference is to use the SageMaker.Client class from the boto3 library. The SageMaker.Client class is a low-level client representing Amazon SageMaker Service. In the following code snippet, we are creating an instance of SageMaker.Client and demonstrating how to access the endpoint using the invoke_endpoint method:

```
import boto3
client = boto3.client("runtime.sagemaker")
# SageMaker Inference endpoint name
endpoint_name = "run_model_prediction"
# Payload for inference which consists of the input data
payload = "..."
# SageMaker endpoint called to get HTTP response (inference)
response = client.invoke_endpoint(
    EndpointName=endpoint_name,
    ContentType="text/csv", # content type
    Body=payload # input data to the endpoint)
```

As shown in the preceding code snippet, the invoke_endpoint method takes in EndpointName (the name of the endpoint; that is, run_model_prediction), ContentType (the type of the input data; that is, "text/csv"), and Body (the input data for model prediction; that is, payload).

In reality, many companies utilize Amazon API Gateway (https://aws.amazon.com/api-gateway) and AWS Lambda (https://aws.amazon.com/lambda) along with SageMaker endpoints, to communicate with the deployed model in a serverless architecture. For the detailed setup, please refer to https://aws.amazon.com/blogs/machine-learning/call-an-amazon-sagemaker-model-endpoint-using-amazon-api-gateway-and-aws-lambda.

Next, we will explain framework-specific approaches to creating an endpoint.

Setting up a TensorFlow inference endpoint

In this section, we will describe a Model class designed specifically for TF – the TensorFlowModel class. Then, we will explain how to use the TF-specific Estimator class for endpoint creation. The complete versions of the code snippets in this section can be found at https://github.com/PacktPublishing/Production-Ready-Applied-Deep-Learning/tree/main/Chapter_9/sagemaker.

Setting up a TensorFlow inference endpoint using the TensorFlowModel class

The TensorFlowModel class is a Model class that is designed for TF models. As shown in the following code snippet, the class can be imported from the sagemaker.tensorflow module and its usage is identical to the Model class:

```
from sagemaker.tensorflow import TensorFlowModel
# Model instance
sm_model = TensorFlowModel(
    model_data=model_s3_path,
    framework_version=tf_framework_version,
    role=role) # IAM role of the notebook
# Predictor
predictor = sm_model.deploy(
    initial_instance_count=1,
    instance_type="ml.c5.xlarge")
```

The constructor of the TensorFlowModel class takes in the same parameters as the constructor of the Model class: the S3 path of the uploaded model (model_s3_path), the TF framework version (Tf_framework_version), and the IAM role for SageMaker (role). In addition, you can provide a Python script for pre- and post-processing the input and output of the model inference by providing entry_point. In this case, the script needs to be named inference. py. For more details, please refer to https://sagemaker.readthedocs.io/en/stable/ frameworks/tensorflow/deploying_tensorflow_serving.html#providing- python-scripts-for-pre-post-processing.

Being a child class of Model, TensorFlowModel also provides a Predictor instance through the deploy method. Its usage is identical to what we described in the preceding section.

Next, you will learn how to deploy your model using the Estimator class, which we have already introduced for the model training on SageMaker in *Chapter 6, Efficient Model Training*.

Setting up a TensorFlow inference endpoint using the Estimator class

As introduced in the *Training a TensorFlow model using SageMaker* section of *Chapter 6, Efficient Model Training*, SageMaker provides the Estimator class, which supports model training on SageMaker. The same class can be used to create and deploy an inference endpoint. In the following code snippet, we are making use of the Estimator class that's been designed for TF, sagemaker.tensorflow. estimator.TensorFlow, to train a TF model and deploy an endpoint using a trained model:

```
from sagemaker.tensorflow.estimator import TensorFlow
# create an estimator
estimator = TensorFlow(
```

```
    entry_point="tf-train.py",
    ...,
    instance_count=1,
    instance_type="ml.c4.xlarge",
    framework_version="2.2",
    py_version="py37" )
# train the model
estimator.fit(inputs)
# deploy the model and returns predictor instance for inference
predictor = estimator.deploy(
    initial_instance_count=1,
    instance_type="ml.c5.xlarge")
```

In the preceding code snippet, the sagemaker.tensorflow.estimator.TensorFlow class takes in the following parameters: entry_point (the script that handles the training; that is, "tf-train.py"), instance_count (the number of instances to use; that is, 1), instance_type (the type of the instance; that is, "ml.c4.xlarge"), framework_version (a PyTorch version; that is, "2.2"), and py_version (a Python version; that is, "py37"). The fit method of the Estimator instance performs the model training. The key method for creating and deploying an endpoint is the deploy method, which creates and hosts an endpoint for the model it trained based on the conditions provided: the initial_instance_count (1) instances of instance_type ("ml.c5.xlarge"). The deploy method of the Estimator class returns a Predictor instance as in the case of the Model class.

In this section, we explained how to create an endpoint for a TF model on SageMaker. In the next section, we will look at how SageMaker supports PyTorch models.

Setting up a PyTorch inference endpoint

This section is designed to cover different ways of creating and hosting an endpoint from a PyTorch model on SageMaker. First, we will introduce a Model class designed for PyTorch models: the PyTorchModel class. Then, we will describe an Estimator class for the PyTorch model. The complete implementations for the code snippets in this section can be found at https://github.com/PacktPublishing/Production-Ready-Applied-Deep-Learning/blob/main/Chapter_9/sagemaker/pytorch-inference.ipynb.

Setting up a PyTorch inference endpoint using the PyTorchModel class

Similar to the TensorFlowModel class, there exists a Model class designed specifically for a PyTorch model, PyTorchModel. It can be instantiated as follows:

```
from sagemaker.pytorch import PyTorchModel
model = PyTorchModel(
```

```
    entry_point="inference.py",
    source_dir="s3://bucket/model",
    role=role, # IAM role for SageMaker
    model_data=pt_model_data, # model file
    framework_version="1.11.0", # PyTorch version
    py_version="py3", # python version
)
```

As shown in the preceding code snippet, the constructor takes in `entry_point`, which defines custom pre- and post-processing logic for the data, `source_dir` (the S3 path of the entry point script), `role` (the IAM role for SageMaker), `model_data` (the S3 path of the model), `framework_version` (the version of PyTorch), and `py_version` (the version of Python).

Since the `PyTorchModel` class inherits the `Model` class, it provides the `deploy` function, which creates and deploys an endpoint, as described in the *Setting up a PyTorch inference endpoint using the Model class* section.

Next, we will introduce an `Estimator` class designed for PyTorch models.

Setting up a PyTorch inference endpoint using the Estimator class

If a trained PyTorch model is not available, the `sagemaker.pytorch.estimator.PyTorch` class can be used to train and deploy a model. The training can be achieved with the `fit` method, as described in the *Training a PyTorch model using SageMaker* section of *Chapter 6, Efficient Model Training*. Being an `Estimator` class, the `sagemaker.pytorch.estimator.PyTorch` class provides the same features as `sagemaker.tensorflow.estimator.TensorFlow`, which we covered in the *Setting up a TensorFlow inference endpoint using the Estimator class* section. In the following code snippet, we are creating an `Estimator` instance for a PyTorch model, training the model, and creating an endpoint:

```
from sagemaker.pytorch.estimator import PyTorch
# create an estimator
estimator = PyTorch(
    entry_point="pytorch-train.py",
    ...,
    instance_count=1,
    instance_type="ml.c4.xlarge",
    framework_version="1.11",
    py_version="py37")
# train the model
estimator.fit(inputs)
```

```
# deploy the model and returns predictor instance for inference
predictor = estimator.deploy(
    initial_instance_count=1,
    instance_type="ml.c5.xlarge")
```

As shown in the preceding code snippet, the constructor of sagemaker.pytorch.estimator. PyTorch takes in the same set of parameters as the Estimator class designed for TF: entry_ point (the script that handles the training; that is, "pytorch-train.py"), instance_count (the number of instances to use; that is, 1), instance_type (the type of the EC2 instance; that is, "ml.c4.xlarge"), framework_version (the PyTorch version; that is, "1.11.0"), and py_version (the Python version; that is, "py37"). The model training (the fit method) and deployment (the deploy method) are achieved the same way as in the previous example in the *Setting up a TensorFlow inference endpoint using the Estimator class* section.

In this section, we covered how to deploy a PyTorch model in two different ways: using the PyTorchModel class and using the Estimator class. Next, we will learn how to create an endpoint for an ONNX model on SageMaker.

Setting up an inference endpoint from an ONNX model

As mentioned in the previous chapter, *Chapter 8, Simplifying Deep Learning Model Deployment*, DL models are often transformed into **open neural network exchange** (**ONNX**) models for deployment. In this section, we will describe how to deploy an ONNX model on SageMaker.

The most standard approach is to use the base Model class. As mentioned in the *Setting up a TensorFlow inference endpoint using the Model class* section, the Model class supports DL models of various types. Fortunately, it provides built-in support for ONNX models as well:

```
from sagemaker.model import Model
# Load an ONNX model file for endpoint creation
sm_model= Model(
    model_data=model_data, # path for an ONNX .tar.gz file
    entry_point="inference.py", # an inference script
    role=role,
    py_version="py3",
    framework="onnx",
    framework_version="1.4.1", # ONNX version
)
# deploy model
predictor = sm_model.deploy(
```

```
    initial_instance_count=1, # number of instances to use
    instance_type=ml.c5.xlarge) # instance type for deploy
```

In the preceding example, we have a trained ONNX model on S3. The key in the `Model` instance creation comes from `framework="onnx"`. We also need to provide an ONNX framework version to `framework_version`. In this example, we are using the ONNX framework version 1.4.0. Everything else is almost identical to the previous examples. Again, the `deploy` function is designed for creating and deploying an endpoint; a `Predictor` instance will be returned for model prediction.

It is also common to use the `TensorFlowModel` and `PyTorchModel` classes for creating an endpoint from an ONNX model. The following code snippet demonstrates such use cases:

```
from sagemaker.tensorflow import TensorFlowModel
# Load ONNX model file as a TensorFlowModel
tf_model = TensorFlowModel(
    model_data=model_data, # path to the ONNX .tar.gz file
    entry_point="tf_inference.py",
    role=role,
    py_version="py3", # Python version
    framework_version="2.1.1", # TensorFlow version
)
from sagemaker.pytorch import PyTorchModel
# Load ONNX model file as a PyTorchModel
pytorch_model = PyTorchModel(
    model_data=model_data, # path to the ONNX .tar.gz file
    entry_point="pytorch_inference.py",
    role=role,
    py_version="py3", # Python version
    framework_version="1.11.0", # PyTorch version
)
```

The preceding code snippets are self-explanatory. Both classes take in a ONNX model path (`model_data`), an inference script (`entry_point`), an IAM role (`role`), a Python version (`py_version`), and versions for each framework (`framework_version`). Like how the `Model` class deploys an endpoint, the `deploy` method will create and host an endpoint from each model.

While endpoints allow us to get the model predictions at any point in time for dynamic input data, there are cases where you need to perform inference on the whole input data stored on an S3 bucket instead of feeding each of them one by one. Therefore, we will look at how we can leverage Batch Transform for this requirement.

Handling prediction requests in batches using Batch Transform

We can use the Batch Transform feature of SageMaker (https://docs.aws.amazon.com/ sagemaker/latest/dg/batch-transform.html) to run inference on a large dataset in one queue. Using the sagemaker.transformer.Transformer class, you can perform model prediction in batches for any dataset on S3 without a persistent endpoint. The details are included in the following code snippet:

```
from sagemaker import transformer
bucket_name = "my-bucket" # S3 bucket with data
# location of the input data
input_location = "s3://{}/{}".format(bucket_name, "input_data")
# location where the predictions will be stored
batch_output = "s3://{}/{}".format(bucket_name, "batch-
results")
# initialize the transformer object
transformer = transformer.Transformer(
    base_transform_job_name="Batch-Transform", # job name
    model_name=model_name, # Name of the inference endpoint
    max_payload= 5, # maximum payload
    instance_count=1, # instance count to start with
    instance_type="ml.c4.xlarge", # ec2 instance type
    output_path=batch_output # S3 for batch inference output)
# triggers the prediction on the whole dataset
tf_transformer = transformer.transformer(
    input_location, # input S3 path for input data
    content_type="text/csv", # input content type as CSV
    split_type="Line" # split type for input as Line)
```

As shown in the preceding code, the sagemaker.transformer.Transformer class takes in base_transformer_job_name (a job name for the transformer job), model_name (the name of the model that holds the inference pipeline), max_payload (the maximum payload in MB allowed), instance_count (the number of EC2 instances to start with), instance_type (the type of EC2 instance), and output_path (an S3 path where the output will be stored). The transformer method will trigger the model prediction on the dataset specified. It takes in the following parameters: input_location (the S3 path where the input data is located), content_type (the content of the input data; that is, "text/csv"), and split_type (this controls how to split the input data; "Line" is used to feed each line of the data as an individual input to the model). In reality, many companies also utilize SageMaker processing jobs (https://docs.aws.amazon.com/

`sagemaker/latest/APIReference/API_ProcessingJob.html`) to perform batch inference, but we will not talk about this in detail.

So far, we have looked at how SageMaker supports hosting an inference endpoint for handling live prediction requests and running model predictions in batches for a static dataset available on S3. In the next section, we will describe how to use **AWS SageMaker Neo** to further improve the inference latency of the deployed model.

Improving SageMaker endpoint performance using AWS SageMaker Neo

In this section, we will explain how SageMaker can further improve the performance of the application by exploiting the underlying hardware resources (EC2 instances or mobile devices). The idea is to compile the trained DL model using **AWS SageMaker Neo** (`https://aws.amazon.com/sagemaker/neo`). After the compilation, the generated Neo model can utilize the underlying device better, thus reducing the inference latency. AWS SageMaker Neo supports models of different frameworks (TF, PyTorch, MxNet, and ONNX) and various types of hardware (OS, chip, architecture, and accelerator). The complete list of supported resources can be found at `https://docs.aws.amazon.com/sagemaker/latest/dg/neo-supported-devices-edge-devices.html`.

Neo model generation can be achieved using the `compile` method of the `Model` class. The `compile` method returns an `Estimator` instance that supports endpoint creation. Let's look at the following example for the details:

```
# sm_model created from Model
sm_model = Model(...)
# instance type of which the model will be optimized for
instance_family = "ml_c5"
# DL framework
framework = "tensorflow"
compilation_job_name = "tf-compile"
compiled_model_path = "s3:..."
# shape of an input data
data_shape = {"inputs":[1, data.shape[0], data.shape[1]]}
estimator = sm_model.compile(
    target_instance_family=instance_family,
    input_shape=data_shape,
    ob_name=compilation_job_name,
    role=role,
    framework=framework,
```

```
    framework_version=tf_framework_version,
    output_path=compiled_model_path)
# deploy the neo model on instances of the target type
predictor = estimator.deploy(
    initial_instance_count=1,
    instance_type=instance_family)
```

In the preceding code, we start with a `Model` instance called `sm_model`. We trigger the `compile` method to compile the loaded model into a Neo model. The following list describes the parameters:

- `target_instance_family`: The EC2 instance type that the model will be optimized for
- `input_shape`: The input data shape
- `job_name`: The name of the compilation job
- `role`: The IAM role of the compiled model output
- `framework`: A DL framework such as TF or PyTorch
- `framework_version`: The version of the framework to use
- `output_path`: The output S3 path where the compiled model will be stored

The `Estimator` instance consists of a `deploy` function that creates the endpoint. The output is a `Predictor` instance that you can use to run the model prediction. In the preceding example, we optimized our model to perform the best on instances of the `ml_c5` type.

Next, we will describe how to integrate the EI accelerator into the endpoints running on SageMaker.

Improving SageMaker endpoint performance using Amazon Elastic Inference

In the *Improving EKS endpoint performance using Amazon Elastic Inference* section, we described how an EI accelerator can reduce the operating cost for an inference endpoint while improving the inference latency by exploiting the available GPU devices. In this section, we will cover EI accelerator integration for SageMaker.

The necessary change is fairly simple; you just need to provide `accelerator_type` when triggering the `deploy` method of a `Model` instance:

```
# deploying a Tensorflow/PyTorch/other model files using EI
predictor = sm_model.deploy(
    initial_instance_count=1, # ec2 initial count
    instance_type="ml.m4.xlarge", # ec2 instance type
    accelerator_type="ml.eia2.medium" # accelerator type)
```

In the preceding code, the `deploy` method creates an endpoint for the given `Model` instance. To attach an EI accelerator to the endpoint, you need to specify the type of accelerator you want (`accelerator_type`) on top of the default parameters (`initial_instance_count` and `instance_type`). For the complete description of using EI for the SageMaker endpoint, please look at `https://docs.aws.amazon.com/sagemaker/latest/dg/ei.html`.

In the following section, we will look at the autoscaling feature of SageMaker, which allows us to handle the changes in the incoming traffic better.

Resizing SageMaker endpoints dynamically using autoscaling

Similar to how the EKS cluster supports autoscaling to automatically scale up or down the endpoints based on the changes in the traffic, SageMaker also provides the autoscaling feature. Configuring autoscaling involves configuring the scaling policy, which defines when the scaling takes place and how many resources are created and destroyed at the time of scaling. The scaling policy for the SageMaker endpoint can be configured from the SageMaker web console. The following steps describe how you can configure autoscaling for the inference endpoints created from a SageMaker notebook:

1. Visit the SageMaker web console, `https://console.aws.amazon.com/sagemaker/`, and click **Endpoints** under **Inference** in the navigation panel on the left-hand side. You may need to provide your credentials to log in.

2. Next, you must choose the endpoint name you want to configure. Under the **Endpoint runtime** settings, choose the model variant that requires the configuration. This feature allows you to deploy multiple versions of a model in a single endpoint, spinning up one container per version. The details on this feature can be found at `https://docs.aws.amazon.com/sagemaker/latest/APIReference/API_runtime_InvokeEndpoint.html`.

3. Under the **Endpoint runtime** settings, select **Configure auto scaling**. This will take you to the **Configure variant automatic scaling** page:

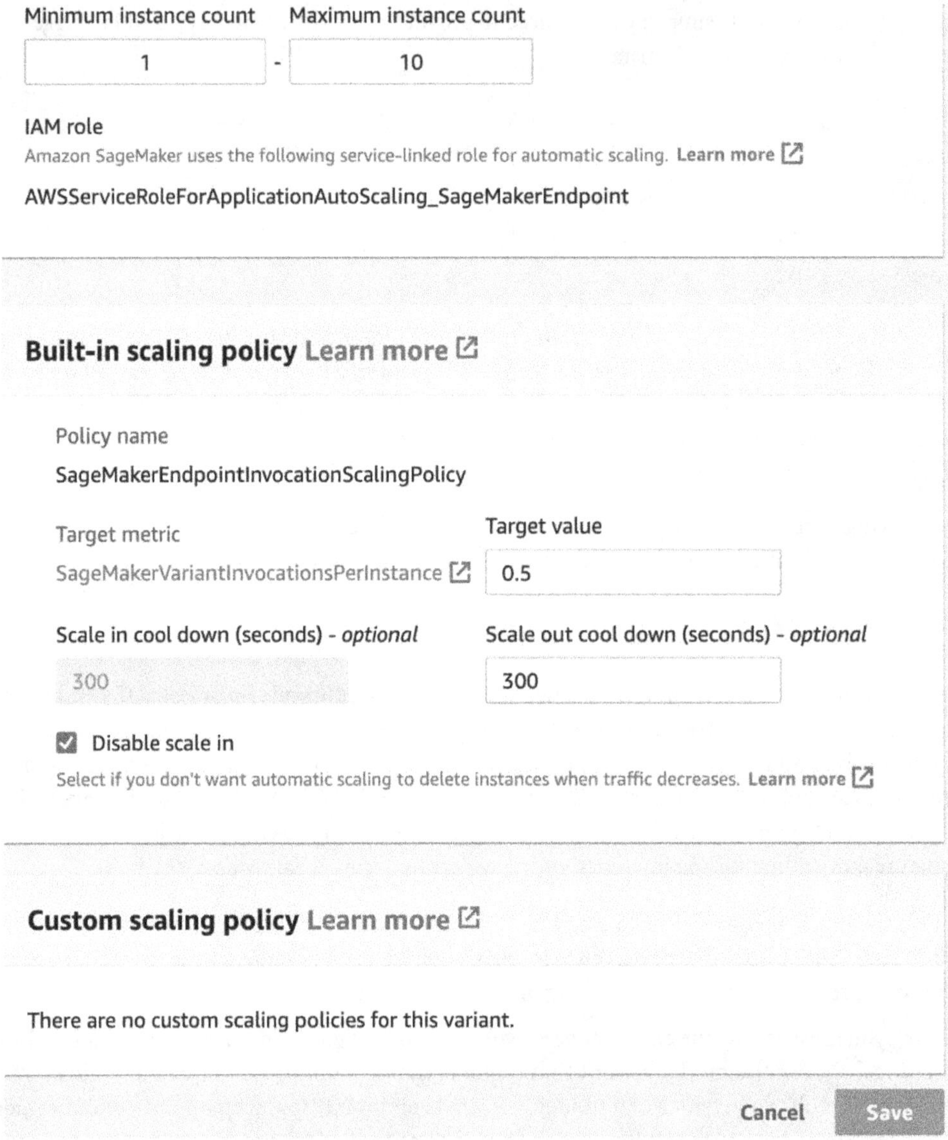

Minimum instance count   Maximum instance count

| 1 | - | 10 |

IAM role

Amazon SageMaker uses the following service-linked role for automatic scaling. **Learn more** ⬀

AWSServiceRoleForApplicationAutoScaling_SageMakerEndpoint

Built-in scaling policy Learn more ⬀

Policy name

SageMakerEndpointInvocationScalingPolicy

Target metric

SageMakerVariantInvocationsPerInstance ⬀

Target value

0.5

Scale in cool down (seconds) - *optional*

300

Scale out cool down (seconds) - *optional*

300

☑ Disable scale in

Select if you don't want automatic scaling to delete instances when traffic decreases. **Learn more** ⬀

Custom scaling policy Learn more ⬀

There are no custom scaling policies for this variant.

Cancel   Save

Figure 9.1 – The Configure variant automatic scaling page of the SageMaker web console

4. Type the minimum number of instances to maintain in the **Minimum instance count** field. The minimum value is 1. This value defines the minimum instance number that will be kept at all times.

5. Type the maximum number of instances of the scaling policy to maintain in the **Maximum instance count** field. This value defines the maximum number of instances allowed at peak traffic.

6.   Fill in the **SageMakerVariantInvocationsPerInstance** field. Each endpoint can have multiple models (or model versions) deployed in a single endpoint hosted across one or more EC2 instances. **SageMakerVariantInvocationsPerInstance** defines the maximum number of invocations allowed per minute for each model variant. This value is used for load balancing. Details on calculating the right number for this field can be found at `https://docs.aws.amazon.com/sagemaker/latest/dg/endpoint-scaling-loadtest.html`.

7.   Fill in the scale-in cooldown and scale-out cooldown. These indicate how long SageMaker will wait before it checks for another round of scaling.

8.   Select the **Disable scale in** checkbox. During an increase in traffic, more instances are started as part of the scale-out process. But these instances can be quickly deleted during the scale-in process if the traffic slows down right after the increase. To avoid a newly created instance from being released as soon as it gets created, this checkbox must be selected.

9.   Click the **Save** button to apply the configuration.

The scaling will be applied to the selected model variant as soon as you click the **Save** button. SageMaker will increase and decrease the number of instances based on the incoming traffic. For more details on auto-scaling, please take a look at `https://docs.aws.amazon.com/autoscaling/ec2/userguide/as-instance-termination.html`.

As the last topic for SageMaker-based endpoints, we will describe how to deploy multiple models through a single endpoint.

Hosting multiple models on a single SageMaker inference endpoint

SageMaker supports deploying multiple models on a single endpoint through **Multimodal Endpoints (MME)**. There are a couple of things you must keep in mind before setting up MME. First, it's recommended to set up multiple endpoints if you want to keep the low latency. Second, the container can only deploy models from the same DL framework. For those who are interested in hosting models from different frameworks, we recommend reading `https://docs.amazonaws.cn/en_us/sagemaker/latest/dg/multi-container-direct.html`. MEE works best when the models are similar in size and expected to perform with similar latencies.

The following steps describe how to set up MME:

1.   Visit the SageMaker web console at `https://console.aws.amazon.com/sagemaker` with your AWS credentials.

2.   Choose **Models** under the **Inference** section of the left navigation panel. Then, click the **Create Model** button at the top right.

3.   Enter a value for the **Model Name** field. This will be used to uniquely identify the target model in the context of SageMaker.

4.   Choose an IAM role with the **AmazonSageMakerFullAccess** IAM policy.

5. Under the **Container definition** section, choose the **Multiple models** option and provide the location of the inference code image and the location of the model artifacts (see *Figure 9.2*):

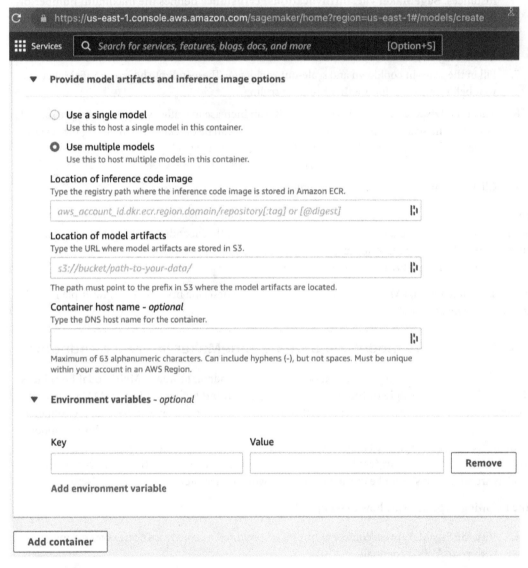

Figure 9.2 – The Multi-modal endpoint configuration page of the SageMaker web console

The former field is used to deploy your models with a custom Docker image (https://docs.aws.amazon.com/sagemaker/latest/dg/your-algorithms-inference-code.html). In this field, you should provide the image registry path where the images are located within Amazon ECR. The latter field specifies the S3 path where the model artifacts reside.

6. Additionally, fill in the **Container host name** field. This specifies details about the host where the inference code image will be created.

7. Choose the **Create Model** button at the end.

Once SageMaker has been configured with MME, we can test the endpoint using `SageMaker.Client` from the `boto3` library as shown in the following code snippet:

```
import boto3
# Sagemaker runtime client instance
runtime_sagemaker_client = boto3.client("sagemaker-runtime")
# send a request to the endpoint targeting  specific model
response = runtime_sagemaker_client.invoke_endpoint(
    EndpointName="<ENDPOINT_NAME>",
    ContentType="text/csv",
    TargetModel="<MODEL_FILENAME>.tar.gz",
    Body=body)
```

In the preceding code, the `invoke_endpoint` function of the `SageMaker.Client` instance sends a request to the created endpoint. The `invoke_endpoint` function takes in `EndpointName` (the name of the created endpoint), `ContentType` (the type of data in the request body), `TargetModel` (the compressed model file in `.tar.gz` format; this is used to specify the target model which the request will be invoking), and `Body` (the input data in `ContentType`). The `response` variable that's returned from the call consists of the prediction results. For the complete description of communicating with the endpoints, please look at `https://docs.aws.amazon.com/sagemaker/latest/dg/invoke-multi-model-endpoint.html`.

Things to remember

a. SageMaker supports endpoint creation through its built-in `Model` class and the `Estimator` class. These classes support models that have been trained with various DL frameworks, including TF, PyTorch, and ONNX. `Model` classes designed specifically for TF and PyTorch frameworks also exist: `TensorFlowModel` and `PyTorchModel`.

b. Once a model has been compiled using AWS SageMaker Neo, the model can exploit the underlying hardware resources better, demonstrating greater inference performance.

c. SageMaker can be configured to use an EI accelerator, reducing the operating cost for inference endpoints while improving the inference latency.

d. SageMaker includes an autoscaling feature that scales the endpoints up and down dynamically based on the volume of incoming traffic.

e. SageMaker supports deploying multiple models on a single endpoint through MME.

Throughout this section, we have described various features that SageMaker provides for deploying a DL model as an inference endpoint.

Summary

In this chapter, we described the two most popular AWS services designed for deploying a DL model as an inference endpoint: EKS and SageMaker. For both options, we started with the simplest setting: creating an inference endpoint from TF, PyTorch, or ONNX models. Then, we explained how to improve the performance of an inference endpoint using the EI accelerator, AWS Neuron, and AWS SageMaker Neo. We also covered how to set up autoscaling to handle the changes in the traffic more effectively. Finally, we discussed the MME feature of SageMaker that is used to host multiple models on a single inference endpoint.

In the next chapter, we will look at various model compression techniques: network quantization, weight sharing, network pruning, knowledge distillation, and network architecture search. These techniques will increase the inference efficiency even further.

10

Improving Inference Efficiency

When a **deep learning** (**DL**) model is deployed on an edge device, inference efficiency is often unsatisfactory. These issues mostly come from the size of the trained network, as it requires a lot of computation. Therefore, many engineers and scientists often sacrifice accuracy for speed when deploying a DL model on an edge device. Furthermore, they focus on reducing the model size as edge devices often have limited storage space.

In this chapter, we will introduce techniques for improving the inference latency while maintaining the original performance as much as possible. First, we will cover **network quantization**, a technique that decreases the network size by using data formats of lower precision for model parameters. Next, we will talk about **weight sharing**, which is also known as weight clustering. It is a very interesting concept where a few model weight values are shared across the whole network, reducing the necessary disk space to store the trained model. We will also talk about **network pruning**, which involves eliminating unnecessary connections within the network. While these three techniques are the most popular, we will also introduce two other interesting subjects: **knowledge distillation** and **network architecture search**. These two techniques achieve model size reduction and inference latency improvement by modifying the network architecture directly during training.

In this chapter, we are going to cover the following main topics:

- Network quantization – reducing the number of bits used for model parameters
- Weight sharing – reducing the number of distinct weight values
- Network pruning – eliminating unnecessary connections within the network
- Knowledge distillation – obtaining a smaller network by mimicking the prediction
- Network Architecture Search – finding the most efficient network architecture

Technical requirements

You can download the supplemental material for this chapter from this book's GitHub repository at `https://github.com/PacktPublishing/Production-Ready-Applied-Deep-Learning/tree/main/Chapter_10`.

Before we deep dive into the individual techniques, we would like to introduce two libraries built on top of **TensorFlow** (**TF**). The first is **TensorFlow Lite** (**TF Lite**), which handles the TF model deployment on mobile, microcontrollers, and other edge devices (`https://www.tensorflow.org/lite`). Some of the techniques we will be describing are only available for TF Lite. The other library is called TensorFlow Model Optimization Toolkit. This library is designed to provide various optimization techniques for TF models (`https://www.tensorflow.org/model_optimization`).

Network quantization – reducing the number of bits used for model parameters

If we look at DL model training in detail, you will notice that the model learns to deal with noisy inputs. In other words, the model tries to construct a generalization for the data it is trained with so that it can generate reasonable predictions even with some noise in the incoming data. Additionally, the DL model ends up using a particular range of numeric values for inference after the training. Following this line of thought, network quantization aims to use simpler representations for these values.

As shown in *Figure 10.1*, network quantization, also called model quantization, is the process of remapping a range of numeric values that the model interacts with to a number system that can be represented with fewer bits – for example, using 8 bits instead of 32 bits to represent a float. Such modifications pose an additional advantage in DL model deployment as edge devices are often missing stable support for arithmetic based on 32-bit floating-point numbers:

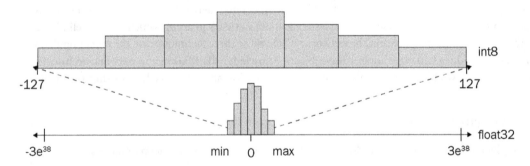

Figure 10.1 – An illustration of the number system remapping from float 32 to int 8 in network quantization

Unfortunately, network quantization involves more than converting a number from high precision into lower precision. This is because DL model inference involves arithmetic that produces numbers with higher precision than the precision of the inputs. In this chapter, we will look at various options in network quantization that overcome the challenge in different ways. If you are interested in learning more about network quantization, we recommend *A Survey of Quantization Methods for Efficient Neural Network Inference*, by Gholami et al.

Network quantization techniques can be categorized into two areas. The first is post-training quantization, while the other is quantization-aware training. The former is designed to quantize a model that has already been trained, while the latter minimizes the accuracy decrease due to quantization process by training a model with lower precision.

Fortunately, these two techniques are both available in standard DL frameworks: TF and PyTorch. In the following sections, we will look at how to perform network quantization in these frameworks.

Performing post-training quantization

First, we will look at how TF and PyTorch support post-training quantization. The modification is simple as it only requires a few additional lines of code. Let's start with TF.

Performing post-training quantization in TensorFlow

By default, a DL model uses floats of 32 bits for the necessary computations and variables. In the following example, we will demonstrate dynamic range quantization where only the fixed parameters (such as weights) are quantized to use 16 bits instead of 32 bits. Please note that you will need to install TF Lite for post-training quantization in TF:

```
import tensorflow as tf
converter = tf.lite.TFLiteConverter.from_saved_model(saved_
model_dir)
converter.optimizations = [tf.lite.Optimize.DEFAULT]
converter.target_spec.supported_types = [tf.float16]
tflite_quant_model = converter.convert()
```

From the quantization, we get a TF Lite model. In the preceding code snippet, we are using the `tf.lite.TFLiteConverter.from_saved_model` function to load a trained TF model and obtain a quantized TF Lite model. Before we trigger the conversion, we need to configure a few things. First, we must set the optimization strategy for quantizing the model weights (`converter.optimizations = [tf.lite.Optimize.DEFAULT]`). Then, we need to specify that we want 16-bit weights from the quantization (`converter.target_spec.supported_types = [tf.float16]`). Actual quantization happens when the `convert` function is triggered. In the preceding code, if we don't specify a 16-bit float type for `supported_types`, we would be quantizing the model to use integers of 8 bits.

Next, we would like to introduce full integer quantization, where every component for the model inference (inputs, activations, as well as weights) is quantized to lower precision. For this type of quantization, you need to provide a representative dataset to estimate the ranges for the activations. Let's look at the following example:

```
import tensorflow as tf
# A set of data for estimating the range of numbers that the
inference requires
representative_dataset = …
```

```
converter = tf.lite.TFLiteConverter.from_saved_model(saved_
model_dir)
converter.optimizations = [tf.lite.Optimize.DEFAULT]
converter.representative_dataset = representative_dataset
converter.target_spec.supported_ops = [tf.lite.OpsSet.TFLITE_
BUILTINS_INT8]
converter.inference_input_type = tf.int8   # or tf.uint8
converter.inference_output_type = tf.int8   # or tf.uint8
tflite_quant_model = converter.convert()
```

The preceding code is almost self-explanatory. Again, we are using the `TFLiteConverter` class for the quantization. First, we configure the optimization strategy (`converter.optimizations = [tf.lite.Optimize.DEFAULT]`) and provide a representative dataset (`converter.representative_dataset = representative_dataset`). Next, we set TF optimizations to be performed in integer representation. Additionally, we need to specify input and output data types by configuring `target_spec`, `inference_input_type`, and `inference_output_type`. Again, the `convert` function in the last line triggers the quantization process.

The two types of post-training quantization in TF are explained thoroughly at `https://www.tensorflow.org/model_optimization/guide/quantization/post_training`.

Next, we will look at how PyTorch achieves post-training quantization.

Performing post-training quantization in PyTorch

In the case of PyTorch, there are two different post-training quantization methods: **dynamic quantization** and **static quantization**. They differ by when the quantization occurs, and have different advantages and disadvantages. In this section, we will provide a high-level description of each algorithm, along with code samples.

Dynamic quantization – quantizing the model at runtime

First, we will look at dynamic quantization, the simplest form of quantization available in PyTorch. This type of algorithm applies the quantization on weights ahead of time while quantization on activations occurs dynamically during inference. Therefore, dynamic quantization is often used in situations where the model execution is mainly throttled by loading weights while computing matrix multiplication is not an issue. This type of quantization is often used for LSTM or Transformer networks.

Given a trained model, dynamic quantization can be achieved as follows. The complete example is available at `https://pytorch.org/tutorials/recipes/recipes/dynamic_quantization.html`:

```
import torch
model = ...
quantized_model = torch.quantization.quantize_dynamic(
```

```
    model,  # the original model
    qconfig_spec={torch.nn.Linear},  # a set of layers to
quantize
    dtype=torch.qint8)  # data type which the quantized tensors
will be
```

To apply dynamic quantization, you need to pass the trained model to the `torch.quantization.quantize_dynamic` function. The other two parameters refer to a set of modules that the quantization will be applied to (`qconfig_spec={torch.nn.Linear}`) and the target data type of the quantized tensors (`dtype=torch.qint8`). In this example, we will quantize the `Linear` layers to use 8-bit integers.

Next, let's look at static quantization.

Static quantization – determining optimal quantization parameters using a representative dataset

The other type of quantization is called static quantization. Like full integer quantization of TF, this type of quantization minimizes the model performance degradation by estimating the range of numbers that the model interacts with using a representative dataset.

Unfortunately, static quantization requires a bit more coding than dynamic quantization. First, you need to insert `torch.quantization.QuantStub` and `torch.quantization.DeQuantStub` operations before and after the network for the necessary tensor conversions, respectively:

```
import torch
# A model with few layers
class OriginalModel(torch.nn.Module):
    def __init__(self):
        super(M, self).__init__()
        # QuantStub converts the incoming floating point
tensors into a quantized tensor
        self.quant = torch.quantization.QuantStub()
        self.linear = torch.nn.Linear(10, 20)
        # DeQuantStub converts the given quantized tensor into
a tensor in floating point
        self.dequant = torch.quantization.DeQuantStub()
    def forward(self, x):
        # using QuantStub and DeQuantStub operations, we can
indicate the region for quantization
        # point to quantized in the quantized model
        x = self.quant(x)
```

```
        x = self.linear(x)
        x = self.dequant(x)
        return x
```

In the preceding network, we have a single `Linear` layer but also have two additional operations initialized in the `__init__` function: `torch.quantization.QuantStub` and `torch.quantization.DeQuantStub`. The former operation is applied to the input tensor to indicate the start of the quantization. The latter operation is applied as the last operation in the `forward` function to indicate the end of the quantization. The following code snippet describes the first step of static quantization – the calibration process:

```
# model is instantiated and trained
model_fp32 = OriginalModel()

...

# Prepare the model for static quantization
model_fp32.eval()
model_fp32.qconfig = torch.quantization.get_default_
qconfig('fbgemm')
model_fp32_prepared = torch.quantization.prepare(model_fp32)
# Determine the best quantization settings by calibrating the
model on a representative dataset.
calibration_dataset = ...
model_fp32_prepared.eval()
for data, label in calibration_dataset:
    model_fp32_prepared(data)
```

The preceding code snippet starts with a trained model, `model_fp32`. To convert the model into an intermediate format for the calibration process, you need to attach a quantization config (`model_fp32.qconfig`) and pass the model to the `torch.quantization.prepare` method. If the model inference runs on a server instance, you must set the `qconfig` property of the model to `torch.quantization.get_default_qconfig('fbgemm')`. If the target environment is a mobile device, you must pass in `'qnnpack'` to the `get_default_qconfig` function. The calibration process can be achieved by passing the representative dataset to the generated model, `model_fp32_prepared`.

The last step is to convert the calibrated model into a quantized model:

```
model_int8 = torch.quantization.convert(model_fp32_prepared)
```

The `torch.quantization.convert` operation in the preceding line of code quantizes the calibrated model (`model_fp32_prepared`) and generates a quantized version of the model (`model_int8`).

Other details on static quantization can be found at `https://pytorch.org/tutorials/advanced/static_quantization_tutorial.html`.

In the next section, we will describe how to perform quantization-aware training in TF and PyTorch.

Performing quantization-aware training

Post-training quantization can reduce the model size significantly. However, it may also reduce the model accuracy significantly. Therefore, the following question arises: can we recover some of the lost accuracy? The answer to this problem might be **quantization-aware training (QAT)**. In this case, the model is quantized before training so that it can learn the generalization directly using the weights and activations of lower precision.

First, let's see how we can achieve this in TF.

Quantization-aware training in TensorFlow

TF provides QAT through TensorFlow Model Optimization Toolkit. The following code snippet describes how you can set up QAT in TF:

```
import tensorflow_model_optimization as tfmot
# A TF model
model = …
q_aware_model = tfmot.quantization.keras.quantize_model(model)
q_aware_model.compile(
               optimizer=...,
               loss=...,
               metrics=['accuracy'])
q_aware_model.fit(...)
```

As you can see, we have used the `tfmot.quantization.keras.quantize_model` function to set up a model for QAT. The output model needs to be compiled using the `compile` function and can be trained using the `fit` function, as in the case of a normal TF model. Surprisingly, this is all you need. The trained model will be already quantized and should provide higher accuracy than the one generated from post-training quantization.

For more details, please refer to the original documentation: `https://www.tensorflow.org/model_optimization/guide/quantization/training_comprehensive_guide`.

Next, we will look at the PyTorch case.

Quantization-aware training in PyTorch

QAT in PyTorch goes through a similar process. Throughout the training process, the necessary calculations are achieved numbers that are clamped and rounded to simulate the effect of quantization. The complete details can be found at `https://pytorch.org/docs/stable/quantization.html#quantization-aware-training-for-static-quantization`. Let's look at how to set up a QAT for PyTorch model.

The setup for QAT is almost identical to what we went through for static quantization in the *Static quantization – determining optimal quantization parameters using a representative dataset* section. The same modification is necessary for the model for both static quantization and QAT; the `torch.quantization.QuantStub` and `torch.quantization.DeQuantStub` operations have to be inserted into the model definition to indicate the region for the quantization. The main difference comes from the intermediate representation of the network since QAT involves updating the model parameters throughout training. The following code snippet describes the difference better:

```
model_fp32 = OriginalModel()
# model must be set to train mode for QAT
model_fp32.train()
model_fp32.qconfig = torch.quantization.get_default_qat_
qconfig('fbgemm')
model_fp32_prepared = torch.quantization.prepare_qat(model_
fp32_fused)
# train the model
for data, label in train_dataset:
    pred = model_fp32_prepared(data)

    ...
# Generate quantized version of the trained model
model_fp32_prepared.eval()
model_int8 = torch.quantization.convert(model_fp32_prepared)
```

In the preceding example, we are using the same network we defined in the *Static quantization – determining optimal quantization parameters using a representative dataset* section: `OriginalModel`. The model should be in `train` mode for QAT (`model_fp32.train()`). Here, we assume that the model will be deployed on a server instance: `torch.quantization.get_default_qat_qconfig('fbgemm')`. In the case of QAT, the intermediate representation of the model is created by passing the original model to the `torch.quantization.prepare_qat` function. You need to train the intermediate representation (`model_fp32_prepared`) instead of the original model (`model_fp32`). Once the training is completed, you can use the `torch.quantization.convert` function to generate the quantized model.

Overall, we have investigated how TF and PyTorch provide QAT to minimize the degradation in model accuracy from the quantization.

> **Things to remember**
>
> a. Network quantization is a simple technique that reduces the inference latency by representing the numbers it deals with in lower precision.
>
> b. There are two types of network quantization: post-training quantization, which applies quantization to a model that is already trained, and QAT, which minimizes the degradation in accuracy by training the model with lower precision.
>
> c. TF and PyTorch support both post-training quantization and QAT with minimal modifications in the training code.

In the next section, we will look at another option for improving inference latency: weight sharing.

Weight sharing – reducing the number of distinct weight values

Weight sharing or **weight clustering** is another technique that can significantly reduce the size of the model. The idea behind this technique is rather simple: let's cluster the weights into groups (or clusters) and use the centroid values instead of individual weight values. In this case, we can store the value of each centroid instead of storing every value for the weights. Therefore, we can compress the model size significantly and possibly speed up the inference process. The key idea behind weight sharing is graphically presented in *Figure 10.2* (adapted from the official TF blog post on weight clustering API: `https://blog.tensorflow.org/2020/08/tensorflow-model-optimization-toolkit-weight-clustering-api.html`):

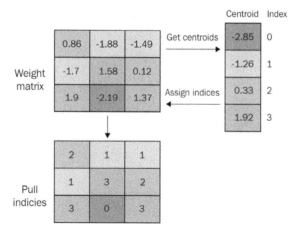

Figure 10.2 – An illustration of weight sharing

Let's learn how to perform weight sharing in TF before looking at how to do the same in PyTorch.

Performing weight sharing in TensorFlow

TF provides weight sharing for both the `Sequential` and `Functional` TF models through TensorFlow Model Optimization Toolkit (`https://www.tensorflow.org/model_optimization/guide/clustering/clustering_example`).

First, you need to define the clustering configuration, as shown in the following code snippet:

```
import tensorflow_model_optimization as tfmot
# A trained model to compress
tf_model = ...
CentroidInitialization = tfmot.clustering.keras.
CentroidInitialization
clustering_params = {
   'number_of_clusters': 10,
   'cluster_centroids_init': CentroidInitialization.LINEAR
}
clustered_model = tfmot.clustering.keras.cluster_weights(tf_
model, **clustering_params)
```

As you can see, weight clustering involves the `tfmot.clustering.keras.cluster_weights` function. We need to provide the trained model (`tf_model`) and a clustering configuration (`clustering_params`). The clustering configuration defines the number of clusters and how each cluster will be initialized. In this example, we are generating 10 clusters that have been initialized using linear centroid initialization (cluster centroids will be evenly spaced between the minimum and maximum values). Other cluster initialization options can be found at `https://www.tensorflow.org/model_optimization/api_docs/python/tfmot/clustering/keras/CentroidInitialization`.

After the model with clustered weights is generated, you can remove all the variables that are not needed during inference using the `tfmot.clustering.keras.strip_clustering` function:

```
final_model = tfmot.clustering.keras.strip_
clustering(clustered_model)
```

Next, we will look at how to perform weight sharing in PyTorch.

Performing weight sharing in PyTorch

Unfortunately, PyTorch does not support weight sharing. Instead, we will provide a high-level description of a possible implementation. In this example, we will try to implement the operation described in *Figure 10.2*. First, we will add a custom function called `cluster_weights` to the model implementation, which you can call after the training for clustering the weights. Then, the forward method will need to be modified slightly, as described in the following code snippet:

```python
from torch.nn import Module
class SampleModel(Module):
# in the case of PyTorch Lighting, we inherit pytorch_
lightning.LightningModule class
    def __init__(self):
        self.layer = …
        self.weights_cluster = … # cluster index for each weight
        self.weights_mapping = … # mapping from a cluster index to
a centroid value
    def forward(self, input):
        if self.training: # in training mode
            output = self.layer(input)
        else: # in eval mode
            # update weights of the self.layer by reassigning each
value based on self.weights_cluster and self.weights_mapping
            output = self.layer(input)
        return output
def cluster_weights(self):
    # cluster weights of the layer
    # construct a mapping from a cluster index to a centroid
value and store at self.weights_mapping
    # find cluster index for each weight value and store at self.
weights_cluster
    # drop the original weights to reduce the model size
# First, we instantiate a model to train
model = SampleModel()
# train the model
…
# perform weight sharing
model.cluster_weights()
model.eval()
```

The preceding code should be self-explanatory as it is pseudocode with comments explaining the key operations. First, the model is trained as if it's a normal model. When the cluster_weights function is triggered, the weights are clustered, and the necessary information for weight sharing is stored within the class; the cluster index for each weight is stored in self.weights_cluster, and the centroid values for each cluster are stored in `self.weights_mapping`. When the model is in `eval` mode, the `forward` operation uses a different set of weights that are constructed from `self.weights_cluster` and `self.weights_mapping`. Additionally, you can add functionality for dropping the existing weights to reduce the model size during deployment. We provide a complete implementation in our repository: `https://github.com/PacktPublishing/Production-Ready-Applied-Deep-Learning/blob/main/Chapter_10/weight_sharing_pytorch.ipynb`.

Things to remember

a. Weight sharing reduces the model size by grouping the distinct weight values and replacing them with the centroid values.

b. TF provides weight sharing through TensorFlow Model Optimization Toolkit, but PyTorch does not provide any support.

Next, let's learn about another popular technique called network pruning.

Network pruning – eliminating unnecessary connections within the network

Network pruning is an optimization process that eliminates unnecessary connections. This technique can be applied after training, but it can also be applied during training where the decrease in model accuracy can be further reduced. With fewer connections, fewer weights are necessary. As a result, we can reduce the model size as well as the inference latency. In the following sections, we will present how to apply network pruning in TF and PyTorch.

Network pruning in TensorFlow

Like model quantization and weight sharing, network pruning for TF is available through TensorFlow Model Optimization Toolkit. Therefore, the first thing you need for network pruning is to import the toolkit with the following line of code:

```
import tensorflow_model_optimization as tfmot
```

To apply network pruning during training, you must modify your model using the `tfmot.sparsity.keras.prune_low_magnitude` function:

```
# data and configurations for training
x_train, y_train, x_text, y_test, x_valid, y_valid,  num_
examples_train, num_examples_test, num_examples_valid  = …
batch_size = ...
```

```
end_step = np.ceil(num_examples_train / batch_size).astype(np.
int32) * epochs
# pruning configuration
pruning_params = {
        'pruning_schedule': tfmot.sparsity.keras.
PolynomialDecay(initial_sparsity=0.3,
final_sparsity=0.5,
begin_step=0,
end_step=end_step)}
#  Prepare a model that will be pruned
model = ...
model_for_pruning = tfmot.sparsity.keras.prune_low_
magnitude(model, **pruning_params)
```

In the preceding code, we have configured network pruning by providing a model and a set of parameters, `pruning_params`, to the `prune_low_magnitude` function. As you can see, we have applied `PolynomialDecay` pruning, which initiates the network at a particular sparsity (`initial_sparsity`) and constructs a network of the target sparsity throughout the training process (https://www.tensorflow.org/model_optimization/api_docs/python/tfmot/sparsity/keras/PolynomialDecay). As shown in the last line, the `prune_low_magnitude` function returns another model that performs network pruning during training.

Before we take a look at the modifications we need to make for the training loop, we would like to introduce another pruning configuration, `tfmot.sparsity.keras.ConstantSparsity` (https://www.tensorflow.org/model_optimization/api_docs/python/tfmot/sparsity/keras/ConstantSparsity). This pruning configuration applies constant sparsity pruning throughout the training process. To apply this type of network pruning, you can simply modify `pruning_params` as shown in the following code snippet:

```
pruning_params = {
        'pruning_schedule': tfmot.sparsity.keras.
ConstantSparsity(0.5, begin_step=0, frequency=100) }
```

As shown in the following code snippet, the training loop requires one additional modification for the callback configurations; we need to use a Keras callback that applies pruning for every optimizer step – that is, `tfmot.sparsity.keras.UpdatePruningStep`:

```
model_for_pruning.compile(...)
callbacks = [tfmot.sparsity.keras.UpdatePruningStep()]
model_for_pruning.fit(x_train, y_train,
```

```
        batch_size=batch_size, epochs=epochs,
        validation_data=(x_valid, y_vallid),
        callbacks=callbacks)
```

The preceding code compiles the model that's been prepared for network pruning and carries out the training. Please keep in mind that the key change comes from the `tfmot.sparsity.keras.UpdatePruningStep` callback specified for the `fit` function.

Finally, you can update the trained model to only remember the sparse weights by passing the model through the `tfmot.sparsity.keras.strip_pruning` function. All `tf.Variable` instances that is not necessary for model inference will be dropped:

```
final_tf_model = tfmot.sparsity.keras.strip_pruning(model_for_
pruning)
```

The presented examples can be directly applied to the `Functional` and `Sequential` TF models. To apply pruning to specific layers or a subset of a model, you need to make the following modifications:

```
def apply_pruning_to_dense(layer):
    if isinstance(layer, tf.keras.layers.Dense):
        return tfmot.sparsity.keras.prune_low_magnitude(layer)
    return layer
model_for_pruning = tf.keras.models.clone_model(model, clone_
function=apply_pruning_to_dense)
```

First, we have defined an `apply_pruning_to_dense` wrapper function that applies the `prune_low_magnitude` function to the target layers. Then, all we need to do is to pass the original model and the `apply_pruning_to_dense` function to the `tf.keras.models.clone_model` function, which generates the new model by running the provided function on the given model.

It is worth mentioning that the `tfmot.sparsity.keras.PrunableLayer` abstract class exists, which is designed for custom network pruning. More details on this can be found at `https://www.tensorflow.org/model_optimization/api_docs/python/tfmot/sparsity/keras/PrunableLayer` and `https://www.tensorflow.org/model_optimization/guide/pruning/comprehensive_guide#custom_training_loop`.

Next, we will look at how pruning can be performed in PyTorch.

Network pruning in PyTorch

PyTorch supports post-training network pruning through the `torch.nn.utils.prune` module. Given a trained network, pruning can be achieved by passing the model to the `global_unstructured` function. Once the model has been pruned, a binary mask is attached, which represents the set of parameters that are pruned. The mask is applied to the target parameter before the `forward` operation, eliminating unnecessary computations. Let's take a look at an example:

```python
# model is instantiated and trained
model = …
parameters_to_prune = (
    (model.conv, 'weight'),
    (model.fc, 'weight')
)
prune.global_unstructured(
    parameters_to_prune,
    pruning_method=prune.L1Unstructured, # L1-norm
    amount=0.2
)
```

As shown in the preceding code snippet, the first parameter of the `global_unstructured` function defines the network components that the pruning will be applied to (`parameters_to_prune`). The second parameter defines the pruning algorithm (`pruning_method`). The last parameter, `amount`, indicates the percentage of parameters to prune. In this example, we are pruning the lowest 20% of the connections based on the L1 norm. If you are interested in other algorithms, you can find the complete list at `https://pytorch.org/docs/stable/nn.html#utilities`.

PyTorch also supports pruning per layer, as well as iterative pruning. You can also define a custom pruning algorithm. The necessary details for the aforementioned functionalities can be found at `https://pytorch.org/tutorials/intermediate/pruning_tutorial.html#pruning-tutorial`.

> **Things to remember**
>
> a. Network pruning is an optimization process that reduces the model size by eliminating unnecessary connections in the network.
>
> b. Both TF and PyTorch support model-level and layer-level network pruning.

In this section, we described how to eliminate unnecessary connections within a network to improve inference latency. In the next section, we will learn about a technique called knowledge distillation, which generates a new model instead of modifying the existing model.

Knowledge distillation – obtaining a smaller network by mimicking the prediction

The idea of knowledge distillation was first introduced in 2015 by Hinton et al. in their publication titled *Distilling the Knowledge in a Neural Network*. In classification problems, Softmax activation is often used as the last operation of the network to represent the confidence for each class as a probability. Since the class with the highest probability is used for the final prediction, the probabilities for the other classes have been considered unimportant. However, the authors believe that they still consist of meaningful information representing how the model interprets the input. For example, if two classes constantly report similar probabilities for multiple samples, the two classes likely have many characteristics in common that makes the distinction between the two difficult. Such information becomes more fruitful when the network is deep because it can extract more information from the data it has seen. Building up from this idea, the authors propose a technique for transferring knowledge of a trained model to a model of a smaller size: knowledge distillation.

The process of knowledge distillation is often referred to as the teacher sharing the knowledge with a student; the original model is referred to as a teacher model, while the smaller model is referred to as a student. As shown in the following diagram, the student model is trained from two different labels constructed from a single input. One label is the ground-truth label, referred to as the hard label. The other label is called the soft label. The soft label is the output probability of the teacher model. The main contribution of the knowledge distillation comes from soft labels filling the missing information in hard labels:

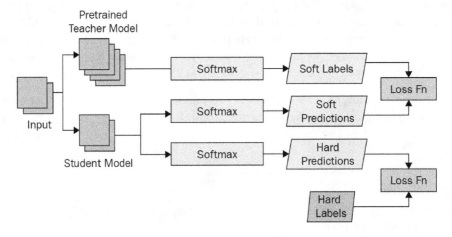

Figure 10.3 – Overview of the knowledge distillation process

From many experiments evaluating the benefit of knowledge distillation, it has been proven that achieving comparable performance using a smaller network is possible. Surprisingly, the simpler network architecture leads to regularization in some cases and results in the student model performing better than the teacher model.

Since the first appearance of this technique, many variations have been introduced. The first set of variations comes from how the knowledge is defined: response-based knowledge (network outputs), feature-based knowledge (intermediate representations), and relation-based knowledge (relationships between layers or data samples). The other set of variations focuses on how to achieve the knowledge transfer: offline distillation (training a student model from a pre-trained teacher model), online distillation (sharing the knowledge as both models get trained), and self-distillation (sharing the knowledge within a single network). We believe that a paper titled *Knowledge distillation: A survey* written by Gou et al. can be a good starting point if you are willing to explore this domain further.

Unfortunately, due to the complexity of the training setup, there isn't a framework that supports knowledge distillation out of the box. However, it can still be a great option if the model network is complex while the output structure is simple.

> **Things to remember**
>
> a. Knowledge distillation is a technique for transferring the knowledge of a trained model to a model of a smaller size.
>
> b. In knowledge distillation, the original model is referred to as a teacher model while the smaller model is referred to as a student. The student model is trained from two labels: ground-truth labels and the output of the teacher model.

Finally, we introduce a technique that modifies the network architecture to reduce the number of model parameters: network architecture search.

Network Architecture Search – finding the most efficient network architecture

Neural Architecture Search (NAS) is the process of finding the best organization of the layers for the given problem. As the search space of the possible network architectures is extremely large, it is not feasible to evaluate every possible network architecture. Therefore, there is a need for a clever way to identify a promising network architecture and evaluate the candidates. Therefore, NAS methods are developed along three different aspects:

- **Search space**: How to construct a search space of a reasonable size
- **Search strategy**: How to explore the search space efficiently

- **Performance estimation strategy**: How to estimate the performance efficiently without training the model completely

Even though NAS is a fast-growing field of research, a few tools are available for TF and PyTorch models:

- Optuna (`https://dzlab.github.io/dltips/en/tensorflow/hyperoptim-optuna`)
- Syne-Tune, which can be used with SageMaker (`https://aws.amazon.com/blogs/machine-learning/run-distributed-hyperparameter-and-neural-architecture-tuning-jobs-with-syne-tune`)
- Katib (`https://www.kubeflow.org/docs/components/katib/hyperparameter`),
- **Neural Network Intelligence (NNI)** (`https://github.com/Microsoft/nni/blob/b6cf7cee0e72671672b7845ab24fcdc5aed9ed48/docs/en_US/GeneralNasInterfaces.md#example-enas-macro-search-space`)
- SigOpt (`https://sigopt.com/blog/simple-neural-architecture-search-nas-intel-sigopt`)

The simplistic version of NAS implementation involves defining a search space from a random layer of organizations. Then, we simply pick the model with the best performance. To reduce the overall search time, we can apply early stopping based on a particular evaluation metric, which will quickly halt the training when the evaluation metric is no longer changing. Such setup reformulates NAS into a hyperparameter tuning problem where the model architecture has become a parameter. We can further improve the search algorithm by applying one of the following techniques:

- Bayesian optimization
- **Reinforcement learning (RL)**
- Gradient-based methods
- Hierarchical-based methods

If you want to explore this space further, we recommend implementing NAS on your own. First, you can exploit the hyperparameter tuning techniques that were introduced in *Chapter 7, Revealing the Secret of Deep Learning Models*. You can start with a random parameter search or a Bayesian optimization approach combined with early stopping. Then, we suggest looking into the RL-based implementation. We also recommend reading a paper called *A Comprehensive Survey of Neural Architecture Search: Challenges and Solutions*, written by Pengzhen Ren et al.

> **Things to remember**
>
> a. NAS is the process of finding the best network architecture for the underlying problem.
>
> b. NAS consists of three components: search space, search strategy, and performance estimation strategy. It involves evaluating networks of different architectures and finding the best one.
>
> c. A few tools for NAS exist: Optuna, Syne-Tune, Katib, NNI, and SigOpt.

In this section, we introduced NAS and how it can generate a network of smaller size.

Summary

In this chapter, we covered a set of techniques that you can use to improve inference latency by reducing the model size. We introduced the three most popular techniques, along with complete examples in TF and PyTorch: network quantization, weight sharing, and network pruning. We also described techniques that reduce the model size by modifying the network architecture directly: knowledge distillation and NAS.

In the next chapter, we will explain how to deploy TF and PyTorch models on mobile devices where the techniques described in this section can be useful.

11
Deep Learning on Mobile Devices

In this chapter, we will introduce how to deploy **deep learning** (**DL**) models developed with **TensorFlow** (**TF**) and **PyTorch** on mobile devices using **TensorFlow Lite** (**TF Lite**) and **PyTorch Mobile**, respectively. First, we will discuss how to convert a TF model into a TF Lite model. Then, we will explain how to convert a PyTorch model into a TorchScript model that PyTorch Mobile can consume. Finally, the last two sections of this chapter will cover how to integrate the converted models into Android and iOS applications (apps).

In this chapter, we're going to cover the following main topics:

- Preparing DL models for mobile devices
- Creating iOS apps with a DL model
- Creating Android apps with a DL model

Preparing DL models for mobile devices

Mobile devices have reshaped how we carry out our daily lives by enabling easy access to the internet; many of our daily tasks heavily depend on mobile devices. Hence, if we can deploy DL models on mobile apps, we should be able to achieve the next level of convenience. Popular use cases include translation among different languages, object detection, and digit recognition, to name a few.

The following screenshots provide some example use cases:

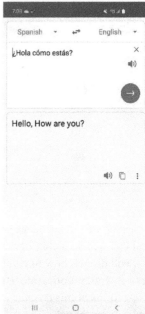

Figure 11.1 – From left to right, the listed apps handle plant identification,
object detection, and machine translation, exploiting the flexibility of DL

There exist many **operating systems (OSs)** for mobile devices. However, two OSs are dominating the mobile market currently: iOS and Android. iOS is the OS for devices from Apple, such as iPhone and iPad. Similarly, Android is the standard OS for devices produced by companies such as—for example—Samsung and Google. In this chapter, we focus on deployments targeted at the two dominating OSs.

Unfortunately, TF and PyTorch models cannot be deployed on mobile devices in their original format. We need to convert them into formats that can run the inference logic on mobile devices. In the case of TF, we need a TF Lite model; we will first discuss how to convert a TF model into a TF Lite model using the `tensorflow` library. PyTorch, on the other hand, involves the PyTorch Mobile framework, which can only consume a TorchScript model. Following TF Lite conversion, we will learn how to convert a PyTorch model into a TorchScript model. Additionally, we will explain how to optimize certain layers of a PyTorch model for the target mobile environment.

It is worth noting that a TF model or a PyTorch model can be converted to **open neural network exchange (ONNX)** Runtime and deployed on mobile (`https://onnxruntime.ai/docs/tutorials/mobile`). Additionally, SageMaker provides built-in support for loading DL models onto edge devices: SageMaker Edge Manager (`https://docs.aws.amazon.com/sagemaker/latest/dg/edge-getting-started-step4.html`).

Generating a TF Lite model

TF Lite is a library used to deploy models on mobile devices, microcontrollers, and other edge devices (https://www.tensorflow.org/lite). A trained TF model needs to be converted into a TF Lite model to be runnable on edge devices. As shown in the following code snippet, the tensorflow library has built-in support for converting a TF model to a TF Lite model (a .tflite file):

```
import tensorflow as tf
# path to the trained TF model
trained_model_dir = "s3://mybucket/tf_model"
# TFLiteConverter class is necessary for the conversion
converter = tf.lite.TFLiteConverter.from_saved_model(trained_
model_dir)
tfl_model = converter.convert()
# save the converted model to TF Lite format
with open('model_name.tflite', 'wb') as f:
   f.write(tfl_model)
```

In the preceding Python code, the from_saved_model function of the tf.lite. TFLiteConverter class loads a trained TF model file. The convert method of this class converts the loaded TF model into a TF Lite model.

As discussed in *Chapter 10, Improving Inference Efficiency*, TF Lite has diverse support for model compression techniques. Popular techniques available from TF Lite include network pruning and network quantization.

Next, let's look at how to convert a PyTorch model into a TorchScript model for PyTorch Mobile.

Generating a TorchScript model

Running a PyTorch model on mobile devices can be achieved using the PyTorch Mobile framework (https://pytorch.org/mobile/home/). Similar to the case of TF, a trained PyTorch model has to be converted into a TorchScript model in order to run the model using PyTorch Mobile (https://pytorch.org/docs/master/jit.html). The main advantage of a torch. jit module developed for TorchScript is the capability of running a PyTorch module outside of the Python environment, such as C++ environment. This is important when deploying a DL model to mobile devices as they do not support Python but support C++. The torch.jit.script method exports the graph of the given DL model into a low-level representation that can be executed in a C++ environment. Complete details on the cross-language support can be found at https://pytorch. org/docs/stable/jit_language_reference.html#language-reference. Please note that TorchScript is still in a beta state.

In order to obtain a TorchScript model from a PyTorch model, you need to pass the trained model to the `torch.jit.script` function, as shown in the following code snippet. The TorchScript model can be further optimized for mobile environments by fusing `Conv2D` and `BatchNorm` layers or removing unnecessary `Dropout` layers using the `optimize_for_mobile` method of the `torch.utils.mobile_optimizer` module (https://pytorch.org/docs/stable/mobile_optimizer.html). Please keep in mind that the `mobile_optimizer` method is also in a beta state:

```
import torch
from torch.utils.mobile_optimizer import optimize_for_mobile
# load a trained PyTorch model
saved_model_file = "model.pt"
model = torch.load(saved_model_file)
# the model should be in evaluate mode for dropout and batch
normalization layers
model.eval()
# convert the model into a TorchScript model and apply
optimization for mobile environment
torchscript_model = torch.jit.script(model)
torchscript_model_optimized = optimize_for_mobile(torchscript_
model)
# save the optimized TorchScript model into a .pt file
torch.jit.save(torchscript_model_optimized, "mobile_optimized.
pt")
```

In the preceding example, we first load the trained model in memory (`torch.load("model.pt")`). The model should be in `eval` mode for the conversion. In the next line, we use the `torch.jit.script` function to convert the PyTorch model into a TorchScript model (`torchscript_model`). The TorchScript model is further optimized for the mobile environment using the `optimize_for_mobile` method; it generates an optimized TorchScript model (`torch_script_model_optimized`). The optimized TorchScript model can be saved as an independent `.pt` file (`mobile_optimized.pt`) using the `torch.jit.save` method.

Things to remember

a. Running a TF model on mobile devices involves the TF Lite framework. The trained model needs to be converted into a TF Lite model. The `TFLiteCoverter` class from the `tensorflow.lite` library is used for the conversion.

b. Running a PyTorch model on a mobile device involves the PyTorch Mobile framework. Given that PyTorch Mobile only supports TorchScript models, the trained model needs to be converted into a `TorchScript` model using torch.jit library.

Next, we will learn how to integrate TF Lite and TorchScript models into an iOS app.

Creating iOS apps with a DL model

In this section, we will cover how to write inference code for TF Lite and TorchScript models for an iOS app. While Swift and Objective-C are the native languages for iOS and can be used together for a single project, we will mainly look at Swift use cases as it is more popular than Objective-C nowadays.

The chapter would be lengthy if we explain every step of iOS app development. Therefore, we relegate the basics to the official tutorial provided by Apple: `https://developer.apple.com/tutorials/app-dev-training`.

Running TF Lite model inference on iOS

In this section, we show how a TF Lite model can be loaded in an iOS app using `TensorFlowLiteSwift`, the native iOS library for TF Lite (`https://github.com/tensorflow/tensorflow/tree/master/tensorflow/lite/swift`). Installing `TensorFlowLiteSwift` can be achieved through CocoaPods, the standard package manager for iOS app development (`https://cocoapods.org`). To download CocoaPods on macOS, you can run the `brew install cocoapods` command on the terminal. Each iOS app development involves a Podfile that lists the libraries that the app development depends on The `TensorFlowLiteSwift` library has to be added to this file, as shown in the following code snippet:

```
pod 'TensorFlowLiteSwift'
```

To install all the libraries in a Podfile, you can run the `pod install` command.

The following steps describe how to load a TF Lite model for your iOS app and run the inference logic. Complete details on the execution can be found at `https://www.tensorflow.org/lite/guide/inference#load_and_run_a_model_in_swift`:

1. The installed libraries can be loaded using the `import` keyword:

    ```
    import TensorFlowLite
    ```

2. Initialize an `Interpreter` class by providing the path to the input TF Lite model:

    ```
    let interpreter = try Interpreter(modelPath: modelPath)
    ```

3. In order to pass the input data to the model, you need to use the `self.interpreter.copy` method to copy the input data into the input `Tensor` object at index 0:

    ```
    let inputData: Data
    inputData = ...
    try self.interpreter.copy(inputData, toInputAt: 0)
    ```

4. Once the input `Tensor` object is ready, the `self.interpreter.invoke` method can be used to run the inference logic:

    ```
    try self.interpreter.invoke()
    ```

5. The generated output can be retrieved using `self.interpreter.output` as a Tensor object that can be further deserialized into an array using the `UnsafeMutableBufferPointer` class:

    ```
    let outputTensor = try self.interpreter.output(at: 0)
    let outputSize = outputTensor.shape.dimensions.reduce(1,
    {x, y in x * y})
    let outputData = UnsafeMutableBufferPointer<Float32>.
    allocate(capacity: outputSize)
    outputTensor.data.copyBytes(to: outputData)
    ```

In this section, we learned how to run TF Lite model inference in an iOS app. Next, we will introduce how to run TorchScript model inference in an iOS app.

Running TorchScript model inference on iOS

In this section, we will learn how to deploy a TorchScript model on an iOS app using PyTorch Mobile. We will start with a Swift code snippet that uses the `TorchModule` module to load a trained TorchScript model. The library you need for PyTorch Mobile is called `LibTorch_Lite`. This library is also available through CocoaPods. All you need to do is to add the following line to the Podfile:

```
pod 'LibTorch_Lite', '~>1.10.0'
```

As described in the last section, you can run the `pod install` command to install the library.

Given a TorchScript model is designed for C++, Swift code cannot run model inference directly. To bridge this gap, there exists the `TorchModule` class, an Objective-C wrapper for `torch::jit::mobile::Module`. To use this functionality in your app, a folder named `TorchBridge` needs to be created under the project and contains `TorchModule.mm` (Objective-C implementation file), `TorchModule.h` (header file), and a bridging header file with the naming convention of a `-Bridging-Header.h` postfix (to allow Swift to load the Objective-C library). The complete sample setup can be found at `https://github.com/pytorch/ios-demo-app/tree/master/HelloWorld/HelloWorld/HelloWorld/TorchBridge`.

Throughout the following steps, we will show how to load a TorchScript model and trigger model prediction:

1. First, you need to import the `TorchModule` class to the project:

    ```
    #include "TorchModule.h"
    ```

2. Next, instantiate `TorchModule` by providing a path to the TorchScript model file:

```
let modelPath = "model_dir/torchscript_model.pt"
let module = TorchModule(modelPath: modelPath)
```

3. The `predict` method of the `TorchModule` class handles the model inference. An input needs to be provided to the `predict` method and the output will be returned. Under the hood, the `predict` method will call the `forward` function of the model through the Objective-C wrapper. The code is illustrated in the following snippet:

```
let inputData: Data
inputData = ...
let outputs = module.predict(input:
UnsafeMutableRawPointer(&inputData))
```

If you are curious about how inference actually works behind the scenes, we recommend that you read the *Run inference* section of `https://pytorch.org/mobile/ios/`.

Things to remember

a. Swift and Objective-C are the standard languages for developing iOS apps. A project can consist of files written in both languages.

b. The `TensorFlowSwift` library is the TF library for Swift. The `Interpreter` class supports TF Lite model inference on iOS.

c. The `LibTorch_Lite` library supports TorchScript model inference on an iOS app through the `TorchModule` class.

Next, we will introduce how to run inference for TF Lite and TorchScript models on Android.

Creating Android apps with a DL model

In this section, we will discuss how Android supports TF Lite and PyTorch Mobile. Java and **Java Virtual Machine (JVM)**-based languages (for example, Kotlin) are the preferred languages for Android apps. In this section, we will be using Java. The basics of Android app development can be found at `https://developer.android.com`.

We first focus on running TF Lite model inference on Android using the `org.tensorflow:tensorflow-lite-support` library. Then, we discuss how to run TorchScript model inference using the `org.pytorch:pytorch_android_lite` library.

Running TF Lite model inference on Android

First, let's look at how to run a TF Lite model on Android using Java. The `org.tensorflow:` `tensorflow-lite-support` library is used to deploy a TF Lite model on an Android app. The library supports Java, C++ (beta), and Swift (beta). A complete list of supported environments can be found at `https://github.com/tensorflow/tflite-support`.

Android app development involves Gradle, a build automation tool that manages dependencies (`https://gradle.org`). Each project will have a `.gradle` file that specifies the project specification in JVM-based languages such as Groovy or Kotlin. In the following code snippet, we list the libraries that the project is dependent on under the `dependencies` section:

```
dependencies {
    implementation 'org.tensorflow:tensorflow-lite-
support:0.3.1'
}
```

In the preceding Gradle code in Groovy, we have specified the `org.tensorflow:tensorflow-lite-support` library as one of our dependencies. A sample Gradle file can be found at `https://docs.gradle.org/current/samples/sample_building_java_applications.html`.

In the following steps, we will look at how to load a TF Lite model and run the inference logic. You can find the complete details about this process at `https://www.tensorflow.org/lite/api_docs/java/org/tensorflow/lite/Interpreter`:

1. The first is to import the `org.tensorflow.lite` library, which contains the `Interpreter` class for TF Lite model inference:

    ```
    import org.tensorflow.lite.Interpreter;
    ```

2. Then, we can instantiate `Interpreter` class by providing a model path:

    ```
    let tensorflowlite_model_path = "tflitemodel.tflite";
    Interpreter = new Interpreter(tensorflowlite_model_path);
    ```

3. The `run` method of the `Interpreter` class instance is used to run the inference logic. It takes in only one `input` instance of type `HashMap` and provides only one `output` instance of `HashMap`:

    ```
    Map<> input = new HashMap<>();
    Input = ...
    Map<> output = new HashMap<>();
    interpreter.run(input, output);
    ```

In the next section, we will learn how to load a TorchScript model into an Android app.

Running TorchScript model inference on Android

In this section, we will explain how to run a TorchScript model in an Android app. To run TorchScript model inference in an Android app, you need a Java wrapper provided by the `org.pytorch:pytorch_android_lite` library. Again, you can specify the necessary library in the `.gradle` file, as shown in the following code snippet:

```
dependencies {
    implementation 'org.pytorch:pytorch_android_lite:1.11'
}
```

Running TorchScript model inference in an Android app can be achieved by following the steps presented next. The key is to use the `Module` class from the `org.pytorch` library, which calls a C++ function for inference behind the scenes (https://pytorch.org/javadoc/1.9.0/org/pytorch/Module.html):

1. First of all, you need to import the `Module` class:

    ```
    import org.pytorch.Module;
    ```

2. The `Module` class provides a `load` function that creates a Module instance by loading the model file provided:

    ```
    let torchscript_model_path = "model_dir/torchscript_
    model.pt";
    Module = Module.load(torchscript_model_path);
    ```

3. The `forward` method of the `Module` instance is used to run the inference logic and generate an output of type `org.pytorch.Tensor`:

    ```
    Tensor outputTensor = module.forward(IValue.
    from(inputTensor)).toTensor();
    ```

While the preceding steps cover basic usage of the `org.pytorch` module, you can find other details from their official documentation: https://pytorch.org/mobile/android.

> **Things to remember**
>
> a. Java and JVM-based languages (for example, Kotlin) are the native languages for Android apps.
>
> b. The `org.tensorflow:tensorflow-lite-support` library is used to deploy a TF Lite model on Android. The `run` method of the `Interpreter` class instance handles model inference.
>
> c. The `org.pytorch:pytorch_android_lite` library is designed for running the TorchScript model in an Android app. The `forward` method from the `Module` class handles the inference logic.

That completes DL model deployment on Android. Now, you should be able to integrate any TF and PyTorch models into an Android app.

Summary

In this chapter, we covered how to integrate TF and PyTorch models into iOS and Android apps. We started the chapter by describing necessary conversions from a TF model to the TF Lite model and from a PyTorch model to the TorchScript model. Next, we provided complete examples for loading TF Lite and TorchScript models and running inference using the loaded models on iOS and Android.

In the next chapter, we will learn how to keep our eyes on the deployed models. We will look at a set of tools developed for model monitoring and describe how to efficiently monitor models deployed on **Amazon Elastic Kubernetes Service (Amazon EKS)** and Amazon SageMaker.

12

Monitoring Deep Learning Endpoints in Production

Due to the difference in development and production settings, it is difficult to assure the performance of **deep learning** (**DL**) models once they are deployed. If any difference exists in model behavior, it must be captured within a reasonable time; otherwise, it can affect downstream applications in negative ways.

In this chapter, our goal is to explain existing solutions for monitoring DL model behavior in production. We will start by clearly describing the benefit of monitoring and what it takes to keep the overall system running in a stable manner. Then, we will discuss popular tools for monitoring DL models and alerting. Out of the various tools we introduce, we will get our hands dirty with **CloudWatch**. We will start with the basics of CloudWatch and discuss how to integrate CloudWatch into endpoints running on **SageMaker** and **Elastic Kubernetes Service** (**EKS**) clusters.

In this chapter, we're going to cover the following main topics:

- Introduction to DL endpoint monitoring in production
- Monitoring using CloudWatch
- Monitoring a SageMaker endpoint using CloudWatch
- Monitoring an EKS endpoint using CloudWatch

Technical requirements

You can download the supplemental material for this chapter from this book's GitHub repository: https://github.com/PacktPublishing/Production-Ready-Applied-Deep-Learning/tree/main/Chapter_12

Introduction to DL endpoint monitoring in production

We will start our chapter by describing the benefits of DL model monitoring for a deployed endpoint. Ideally, we should analyze information related to incoming data, outgoing data, model metrics, and traffic. A system that monitors the listed data can provide us with the following benefits.

Firstly, *the input and output information for the model can be persisted in a data storage solution (for example, a Simple Storage Service (S3) bucket) for understanding data distributions*. Detailed analysis of the incoming data and predictions can help in identifying potential improvements for the downstream process. For example, monitoring the incoming data can help us in identifying bias in model predictions. Models can be biased toward specific feature groups while handling incoming requests. This information can guide us on what we should be considering when we are training a new model for the following deployment. Another benefit comes from the model's explainability. The reasoning behind a model predictions needs to be explained for business purposes or legal purposes. This involves the techniques we have described in *Chapter 9, Scaling a Deep Learning Pipeline*.

Another key metric we should be tracking is the **throughput** of the endpoint, which can help us improve user satisfaction. *A model's behavior may change depending on the volume of incoming requests and the computational power of the underlying machines*. We can monitor inference latency with respect to incoming traffic to build stable and efficient inference endpoints for users.

At a high level, monitoring for DL models can be categorized into two areas: **endpoint monitoring** and **model monitoring**. In the former area, we aim to collect data related to endpoint latency and throughput of the target endpoint. The latter area is focused on improving model performance; we need to collect incoming data, predictions, and model performances, as well as inference latency. While many use cases of model monitoring are achieved in an online fashion on a running endpoint, it is also applied during the training and validation process in an offline fashion with the goal of understanding the model's behavior prior to deployment.

In the following section, we will introduce popular tools for monitoring DL models.

Exploring tools for monitoring

Tools for monitoring can be mostly categorized into two groups, depending on what they are designed for: **monitoring tools** and **alerting tools**. Covering all tools explicitly is out of the scope of this book; however, we will introduce a few of them briefly to explain the benefits that monitoring and alerting tools aim to provide. Please note that the boundary is often unclear, and some tools may be built to support both features.

Let's dive into monitoring tools first.

Prometheus

Prometheus is an open-source monitoring and alerting tool (https://prometheus.io). Prometheus stores data delivered from the application in local storage. It uses a time-series database for storing, aggregating, and retrieving metrics, which aligns well with the nature of monitoring tasks. Interacting with Prometheus involves using **Prometheus Query Language (PromQL)** (`https://prometheus.io/docs/prometheus/latest/querying/basics`). Prometheus is designed to process metrics such as **central processing unit (CPU)** usage, memory usage, and latency. Additionally, custom metrics such as model performance or distribution of incoming and outgoing data can be ingested for monitoring.

CloudWatch

CloudWatch is a monitoring and observability service designed by **Amazon Web Services (AWS)** (`https://aws.amazon.com/cloudwatch`). CloudWatch is easy to set up compared to setting up a dedicated Prometheus service, as it handles data storage management behind the scenes. By default, most AWS services such as AWS Lambda and EKS clusters use CloudWatch to persist metrics for further analysis. Also, CloudWatch can support alerting users through emails or Slack messages for unusual changes from the monitored metric. For example, you can set a threshold for a metric and get notified if it goes above or below the predefined threshold. Details of the alerting feature can be found at `https://docs.aws.amazon.com/AmazonCloudWatch/latest/monitoring/AlarmThatSendsEmail.html`.

Grafana

Grafana is a popular tool designed for visualizing metrics collected from monitoring tools (`https://grafana.com`). Metrics data from CloudWatch or AWS-managed Prometheus can be read by Grafana for visualization. For a complete description of these configurations, we recommend you to read `https://grafana.com/docs/grafana/latest/datasources/aws-cloudwatch` and `https://docs.aws.amazon.com/prometheus/latest/userguide/AMP-onboard-query-standalone-grafana.html`.

Datadog

One of the popular proprietary solutions is **Datadog** (`https://www.datadoghq.com`). This tool provides a wide variety of monitoring features: log monitoring, application performance monitoring, network traffic monitoring, and real-time user monitoring.

SageMaker Clarify

SageMaker has a built-in support for monitoring endpoints created from SageMaker, **SageMaker Clarify** (`https://aws.amazon.com/sagemaker/clarify`). SageMaker Clarify comes with a **software development kit (SDK)** which helps understand the performance of the model and its bias in predictions. Details of SageMaker Clarify can be found at `https://docs.aws.amazon.com/sagemaker/latest/dg/model-monitor.html`.

In the following section, we will introduce alerting tools.

Exploring tools for alerting

An *incident* is an event that requires a follow-up action, such as a failed job or a build. While monitoring tools can capture unusual changes, they often lack incident management and automation for the responding process. Alerting tools close this gap by providing many of these features out of the box. Therefore, many companies often integrate explicit alerting tools to respond to incidents in a timely manner.

In this section, we will introduce the two most popular alerting tools: PagerDuty and Dynatrace.

PagerDuty

As a tool for alerting and managing the **incident response** (**IR**) process, many companies integrate **PagerDuty** (https://www.pagerduty.com). On top of the basic alerting feature, PagerDuty supports assigning priorities to incidents based on their type and severity. PagerDuty can read data from several popular monitoring software such as Prometheus and Datadog (https://aws.amazon.com/blogs/mt/using-amazon-managed-service-for-prometheus-alert-manager-to-receive-alerts-with-pagerduty). It can also be integrated with CloudWatch with minimal code changes (https://support.pagerduty.com/docs/aws-cloudwatch-integration-guide).

Dynatrace

Dynatrace is another proprietary tool for monitoring entire clusters or networks and alerting incidents (https://www.dynatrace.com). Information related to resource usage, traffic, and response time of running processes can be easily monitored. Dynatrace has a unique alerting system based on alerting profiles. These profiles define how the system delivers notifications across the organization. Dynatrace has built-in push notifications, but it can be integrated with other systems that provide a notification feature, such as Slack and PagerDuty.

> **Things to remember**
>
> a. Monitoring information related to incoming data, outgoing data, model metrics, and traffic volumes for an endpoint allows us to understand the behavior of our endpoint and helps us in identifying potential improvements.
>
> b. Prometheus is an open sourced monitoring and alerting system that can be used for monitoring metrics of a DL endpoint. CloudWatch is a monitoring service from AWS designed for logging a set of data and tracking unusual changes from incoming and outgoing traffic.
>
> c. PagerDuty is a popular alerting tool that handles the complete life cycle of an incident.

In this section, we looked at why we need monitoring for a DL endpoint and provided a list of tools available. In the remaining sections of this chapter, we will look in detail at CloudWatch, the most common monitoring tool, as it is integrated well into most services within AWS (for example, SageMaker).

Monitoring using CloudWatch

First, we will introduce a few key concepts in CloudWatch: logs, metrics, alarms, and dashboards. **CloudWatch** persists ingested data in the form of logs or metrics organized by timestamps. As the name suggests, *logs* refer to text data emitted throughout the lifetime of a program. On the other hand, *metrics* represent organized numeric data such as CPU or memory utilization. Since metrics are stored in an organized matter, CloudWatch supports aggregating metrics and creating histograms from collected data. An *alarm* can be set up to alert if unusual changes are reported for the target metric. Also, a *dashboard* can be set up to get an intuitive view of selected metrics and raised alarms.

In the following example, we will describe how to log metric data using a CloudWatch service client from the boto3 library. The metric data is structured as a dictionary and consists of metric names, dimensions, and values. The idea of dimensions is to capture factual information about the metric. For example, a metric name city can have a value of New York City. Then, dimensions can capture specific information such as hourly counts of fire accidents or burglaries:

```python
import boto3
# create CloudWatch client using boto3 library
cloudwatch = boto3.client('cloudwatch')
# metrics data to ingest
data_metrics=[
    {
        'MetricName': 'gross_merchandise_value',
        'Dimensions': [
            {
                'Name': 'num_goods_sold',
                'Value': '369'
            } ],
        'Unit': 'None',
        'Value': 900000.0
    } ]
# ingest the data for monitoring
cloudwatch.put_metric_data(
    MetricData=data_metrics, # data for metrics
    Namespace='ECOMMERCE/Revenue' # namespace to separate
domain/projects)
```

In the preceding code snippet, we first create a cloudwatch service client for CloudWatch using the boto3.client function. This instance will allow us to communicate with CloudWatch from a Python environment. The key method for logging a set of data is `put_metric_data`. This function `put_metric_data` method from the CloudWatch client instance takes in `MetricData` (the target metric data to ingest into CloudWatch: `data_metrics`) and `Namespace` (container for the metric data: `'ECOMMERCE/Revenue'`). Data from different namespaces is managed separately to support efficient aggregation.

In this example, the `data_metrics` metric data contains a field `MetricName` of `gross_merchandise_value` with the value of `900000.0`. The unit for `gross_merchandise_value` is defined as `None`. Additionally, we are providing the number of goods sold (`num_goods_sold`) as additional dimension information.

For a complete description of CloudWatch concepts, please refer to `https://docs.aws.amazon.com/AmazonCloudWatch/latest/monitoring/cloudwatch_concepts.html`.

> **Things to remember**
>
> a. CloudWatch persists ingested data in the form of logs or metrics organized by timestamps. It supports setting up an alarm for unusual changes and provides effective visualization through dashboards.
>
> b. Logging a metric to CloudWatch can be easily achieved using the `boto3` library. It provides a service client for CloudWatch that supports logging through the `put_metric_data` function.

While logging for CloudWatch can be done explicitly as described in this section, SageMaker provides built-in logging features for some of the out-of-the-box metrics. Let's take a closer look at them.

Monitoring a SageMaker endpoint using CloudWatch

Being an end-to-end service for **machine learning**, SageMaker is one of the main tools that we use to implement various steps of a DL project. In this section, we will describe the last missing piece: monitoring an endpoint created with SageMaker. First, we will explain how you can set up CloudWatch-based monitoring for training where metrics are reported in batches offline. Next, we will discuss how to monitor a live endpoint.

The code snippets in this section are designed to run on SageMaker Studio. Therefore, we first need to define an AWS **Identity and Access Management** (**IAM**) role and a session object. Let's have a look at the first code snippet:

```
import sagemaker
# IAM role of the notebook
role_exec=sagemaker.get_execution_role()
# a sagemaker session object
sag_sess=sagemaker.session()
```

In the preceding code snippet, the `get_execution_role` function provides the IAM role for the notebook. `role_exec.sagemaker.session` provides a SageMaker `sag_sess` SageMaker session object required for the job configuration.

Monitoring a model throughout the training process in SageMaker

The logging during model training involves SageMaker's `Estimator` class. It can process printed messages using `regex` expressions and store them as metrics. You can see an example here:

```
import sagemaker
from sagemaker.estimator import Estimator
# regex pattern for capturing error metrics
reg_pattern_metrics=[
    {'Name':'train:error','Regex':'Train_error=(.*?);'},
    {'Name':'validation:error','Regex':'Valid_error=(.*?)'}]
# Estimator instance for model training
estimator = Estimator(
    image_uri=...,
    role=role_exec,
    sagemaker_session=sag_sess,
    instance_count=...,
    instance_type=...,
    metric_definitions=reg_pattern_metrics)
```

In the preceding code snippet, we create `estimator`, which is an `Estimator` instance for training. Explanations for most of the parameters can be found in *Chapter 6, Efficient Model Training*. The additional parameter we are defining in this example is `metric_definitions`. We are passing in `reg_pattern_metrics`, which defines a set of **regular expressions** (**regex**) search patterns put `Train_error=(.*?)` and `Valid_error=(.*?)`, training and evaluation logs. Texts that match the given patterns will be persisted as metrics in CloudWatch. For the complete details of offline metrics recording throughout model training using the `Estimator` class, please refer to `https://docs.aws.amazon.com/sagemaker/latest/dg/training-metrics.html`. We want to mention that specific training job metrics (such as memory, CPU, **graphics processing unit** (**GPU**), and disk utilization) are automatically logged, and you can monitor them either through CloudWatch or SageMaker console.

Monitoring a live inference endpoint from SageMaker

In this section, we will describe SageMaker's CloudWatch-based monitoring feature for an endpoint. In the following code snippet, we are presenting a sample `inference.py` script with an `output_handler` function. This file is assigned for an `entry_point` parameter of SageMaker's `Model` or `Estimator` class to define additional pre- and postprocessing logic. Details of `inference.py` can be found in *Chapter 9, Scaling a Deep Learning Pipeline*. The `output_handler` function is designed to process model predictions and log metric data using the `print` function. The printed messages get stored as logs in CloudWatch:

```python
# inference.py
def output_handler(data, context):
    # retrieve the predictions
    results=data.content
    # data that will be ingested to CloudWatch
    data_metrics=[
        {
            'MetricName': 'model_name',
            'Dimensions': [
                {
                    'Name': 'classify',
                    'Value': results
                } ],
            'Unit': 'None',
            'Value': "classify_applicant_risk"
        } ]
    # print will ingest information into CloudWatch
    print(data_metrics)
```

In the preceding inference code, we first get a model prediction (`results`) and construct a dictionary for metric data (`data_metrics`). The dictionary already has a `MetricName` value of `model_name` and a dimension named `classify`. The model prediction will be specified for the `classify` dimension. SageMaker will collect printed metric data and ingest it to CloudWatch. A sample approach to continuous model monitoring for quality drift is described online at https://sagemaker-examples.readthedocs.io/en/latest/sagemaker_model_monitor/model_quality/model_quality_churn_sdk.html. This page nicely explains how you can leverage CloudWatch in such scenarios.

> **Things to remember**
>
> a. The `Estimator` class from SageMaker provides built-in support for CloudWatch-based monitoring during training. You need to pass a set of regex patterns to the `metric_definitions` parameter when constructing an instance.
>
> b. Printed messages from a SageMaker endpoint get stored as CloudWatch logs. Therefore, we can achieve monitoring by logging metric data through an `entry_point` script.

In this section, we explained how SageMaker supports CloudWatch-based monitoring. Let's look at how EKS supports monitoring for inference endpoints.

Monitoring an EKS endpoint using CloudWatch

Along with SageMaker, we have described EKS-based endpoints in *Chapter 9, Scaling a Deep Learning Pipeline*. In this section, we describe CloudWatch-based monitoring available for EKS. First, we will learn how EKS metrics from the container can be logged for monitoring. Next, we will explain how to log model-related metrics from an EKS inference endpoint.

Let's first look at how to set up CloudWatch for monitoring an EKS cluster. The simplest approach is to install a CloudWatch agent in the container. Additionally, you can install **Fluent Bit**, an open source tool that further enhances the logging process (`www.fluentbit.io`). For a complete explanation of CloudWatch agents and Fluent Bit, please read `https://docs.aws.amazon.com/AmazonCloudWatch/latest/monitoring/Container-Insights-setup-EKS-quickstart.html`.

Another option is to persist the default metrics sent by the EKS control plane. This can be easily enabled from the EKS web console (`https://docs.aws.amazon.com/eks/latest/userguide/control-plane-logs.html`). The complete list of metrics emitted from the EKS control plane can be found at `https://aws.github.io/aws-eks-best-practices/reliability/docs/controlplane`. For example, if you are interested in logging latency-related metrics, you can use `apiserver_request_duration_seconds*`.

To log model-related metrics during model inference, you need to instantiate `boto3`'s CloudWatch service client within the code and log them explicitly. The code snippet included in the previous section, *Monitoring a SageMaker endpoint using CloudWatch*, should be a good starting point.

> **Things to remember**
>
> a. Logging endpoint-related metrics from an EKS cluster can be achieved by using a CloudWatch agent or persisting default metrics sent by the EKS control plane.
>
> b. Model-related metrics need to be logged explicitly using the `boto3` library.

As the last topic of this section, we explained how to log various metrics to CloudWatch from an EKS cluster.

Summary

Our goal in this chapter was to explain why you need to monitor an endpoint running a DL model and to introduce popular tools in this domain. The tools we introduced in this chapter are designed for monitoring a set of information from an endpoint and alerting an incident when there are sudden changes to the monitored metrics. The tools that we covered are CloudWatch, Prometheus, Grafana, Datadog, SageMaker Clarify, PagerDuty, and Dynatrace. For completeness, we looked at how CloudWatch can be integrated into SageMaker and EKS for monitoring an endpoint as well as model performance.

In the next chapter, as the last chapter of this book, we will explore the process of evaluating a completed project and discussing potential improvements.

13
Reviewing the Completed Deep Learning Project

The last phase of a **deep learning** (**DL**) project is the reviewing process. During the planning phase, the responsibility of each stakeholder has been defined and the goal of the project has been set. In this phase, stakeholders must group again to revisit the responsibilities and the goal to evaluate whether the project was carried out as planned or not. Such a process can be summarized as a **post-implementation review** (**PIR**). To guide the reviewing process further, we also describe different ways of evaluating a completed project, including but not limited to gap analysis, estimate at completion, and sustainability analysis. In addition to project evaluation, the details of the project need to be recorded and potential improvements must be discussed so that the next project can be achieved more efficiently.

In this chapter, we are going to cover the following main topics:

- Reviewing a DL project
- Gathering reusable knowledge, concepts, and artifacts for future projects

Reviewing a DL project

Post-implementation review (**PIR**) is the process of revisiting how the project was carried out. Throughout this process, you will compare the final state of the project against the goal state and organize the generated artifacts from the current project for reusability. Overall, this process should lead you to a broad understanding of the project's success or failure. Furthermore, it will give you a clear indication of how future projects should be managed and how to avoid the mistakes made in the current project. Following this line of thought, you should have a bigger picture in mind all the time beyond the scope of the current project; one project might be completed, but the insights that you obtained will be reusable for future projects.

Conducting a post-implementation review

The PIR process consists of the following steps. Please keep in mind that the process can start before the final deployment of the deliverable:

1. First, try to answer the following key questions: has the project been completed using the available budget and in time? Was it successful?

2. Revisit **key performance indicators** (**KPI**) and other metrics that were defined in the initial phase of your project. List out the metrics achieved and think about where the improvements can be made. You can refer to *Chapter 1, Effective Planning of Deep Learning-Driven Projects*, for the details on various evaluation metrics.

3. Perform GAP analysis (`https://www.batimes.com/articles/do-we-need-a-mature-gap-analysis`). It is a good starting point to get a detailed perspective on project performance. The GAP approach compares the actual performance (of the deployed DL system) against the target performance (defined in the planning phase of the project) across all the objectives. The comparison should lead to possible improvements and additional optimizations.

4. Document opinions from stakeholders and their perspectives on possible improvements. Try to understand stakeholders' satisfaction levels on the project completion.

5. Prepare a detailed cost analysis that summarizes the money spent versus the allocated budget. Each analysis must be linked to each step: development, deployment, monitoring, and maintenance. Try to find the places where the initial estimations were incorrect and think about how you can plan the details better in the next project.

6. Review the steps taken for each task and understand the bottlenecks. Try to identify the places where your process wasn't perfect and discuss how it can be avoided in future projects.

7. Compose a short document summarizing all the points that we just described and have it evaluated by the stakeholders. Focus on whether the project has successfully achieved the initial objective but also remember that the goal of PIR is not only to show how successful your project was. It's important that the participants share what they learned throughout the project as well.

8. Make sure the PIR documentation is accessible by anyone within the organization as a reference for future projects.

The key item in the PIR process is to evaluate the true value of the project. We will introduce various techniques for effective project evaluation in the next section.

Understanding the true value of the project

Let's look at a few aspects that you should keep in mind in this final stage. First, you need to revisit the due dates and estimated resource usage defined in the planning phase. These two factors should've affected the spending directly throughout the project. Even if your project goes beyond the allocated budget or timeline, the project can be considered successful if the return is greater than the resources put in. For a project, **return on investment** (**ROI**) can be calculated using a simple formula:

ROI = [(Financial Value - Project Cost) / Project Cost] x 100

The comparison between the anticipated ROI (calculated before the project or based on the initial estimations) and the actual ROI should give you an additional angle for the project evaluation.

We will not cover all the performance measures that can be used to indicate the final status of the project as there can be many of them: ROI, revenue growth, revenue per customer, profit margin, cost of quality, schedule performance, customer satisfaction, customer retention rate, productivity, level of alignment with business goals, and so on (`https://financesonline.com/10-project-management-success-metrics-to-measure-your-team-performance`). However, we would like to call out **estimate at completion** (**EAC**), the metric that can be used for every phase of the project. It is used to predict the total cost of the project. Comparing EAC against the initially estimated budget at completion, you will be able to review whether you are on track with the initial cost estimation. Along with EAC, it is recommended to track the expenses throughout the project and the cost variance for each activity. Overall, the findings from this process will help you to minimize expenses and increase profits.

A project management standard, **Project Management Body of Knowledge** (**PMBOK**), prioritizes **planned value** (**PV**), **earned value** (**EV**), and **actual cost** (**AC**) as the three crucial metrics for measuring project performance (`https://projectmanagementacademy.net`):

- PV, also called the **budgeted cost of work scheduled** (**BCWS**), is just a cost estimation of the planned activities at any given time. It is mainly used for the baseline.

- EV is called the **budgeted cost of work performed** (**BCWP**) and is the sum of the budget for the activities accomplished during a period of time. The comparison between EV and PV will indicate whether you are on track with resource usage or not.

- AC is also referred to as the **actual cost of work performed** (**ACWP**) and is the total cost of the work performed. Tracking AC and comparing it against the planned spending will help you to understand whether you are on the right path to a successful completion of the project within the budget.

In DL projects, we commonly evaluate both model-related metrics (such as precision, recall, and f1-score) and business-related metrics (such as conversion rate, click-through rate, lifetime value, user engagement measure, and savings in operational cost). Therefore, the definition of the key objective might be more complex than the one for non-DL projects.

Apart from the aspects we have covered so far, you should also consider the **sustainability** of the project. By reviewing the sustainability, you will understand whether your project fulfills the objectives without making negative impacts on the pillars of sustainability: economic, environmental, social, and administrative.

> **Things to remember**
>
> a. The last step of the project is to understand how the project is carried out and discuss how it can be achieved more efficiently in the future.
>
> b. Throughout PIR, you need to review every phase of the project and obtain a broad understanding of the project's success or failure.
>
> c. KPI analysis, GAP analysis, cost analysis, benchmark comparison, and ROI calculations are extremely useful when evaluating the overall project.

Next, we will look at how to efficiently organize and share the collected know-how from the PIR process.

Gathering the reusable knowledge, concepts, and artifacts for future projects

Your DL projects will result in many artifacts that can be reused in the future. For example, the processed data used during the model training can be reused for other analytical tasks, the model implementation can be adapted to other applications, and the infrastructure set up for monitoring tasks can be reconfigured for different projects. To be able to reuse these artifacts, you need to archive them correctly and ensure that sufficient documentation exists. Let's have a look at some procedures that you can implement to make your life easier in this process:

1. Set up versioning standards for development environments, data, implementations, and models. They should be defined at the early stage of the project, and all the team members should follow them:

    *   Add versioning for the code base using Git (https://git-scm.com). The project can be linked with GitHub (https://github.com), GitLab (https://gitlab.com), or AWS CodeCommit (https://aws.amazon.com/codecommit) for better management of the code base.

    *   Set up versioning for the data and model. Details can be found in *Chapter 4, Experiment Tracking, Model Management, and Dataset Versioning*.

    *   Keep separate documentation for each of the project stages, environments, and resources in easily accessible spaces such as **Confluence**, **SharePoint**, **Google Drive**, or **Asana**.

2. Introduce standards for programming and documentation. Ensure that the standards are followed throughout code reviews. Remember to always log details about development environments and crucial library dependencies. Utilize virtual environment tools such as Docker or Anaconda to keep them in a reproducible manner.

3. Summarize the key concepts that are used repeatedly in the project and make sure they are thoroughly documented and easily accessible.

4. Continuously review the status of the stored information to make sure that they are kept up to date throughout the project execution.

In the final stage of your project, it is recommended that the artifacts are reviewed one more time to fill in the missing details. Please keep in mind that your next project can be achieved more efficiently if these resources can handle many of the repetitive tasks.

Depending on your geographical location, you may need to follow specific laws when the deliverable consumes sensitive data. Common ones include GDPR, HIPAA, FCRA, FERPA, GLBA, ECPA, COPPA, and VPPA (`https://www.nytimes.com/wirecutter/blog/state-of-privacy-laws-in-us/`). This final phase of the project would be a good time to ensure that all the regulations and compliance procedures are followed prior to any audits from external organizations.

Things to remember

a. In the final stage of the project, you need to review all the artifacts generated from the project. Their organization and documentation must be revisited so that they can be easily retrieved in the future.

b. Setting up a process for artifact management will allow you to keep them organized. For example, defining standards for the documentation and following them will help you in keeping the resources in a consistent manner.

Summary

You have reached the final phase of your DL project. In this chapter, we described the steps you need to follow to wrap up the project. We first described how to apply PIR to evaluate the project and understand the potential improvements. In this phase, you also need to make sure the artifacts generated from the project are organized and thoroughly documented so that they can be reused for the next project. Lastly, we would like to mention that celebration is another key component of a DL project. All the stakeholders have put in their efforts to carry out the project. You must spend some time thanking all the team members and applauding their achievements.

Throughout this book, you have learned how to carry out a DL project at a high standard. Starting from the basic concepts in DL, we have described each phase of a DL project thoroughly, along with various tools you can use to carry out the task at hand efficiently. The book emphasizes scalability and explains how you can achieve data processing and model training using various cloud services. Overall, you are now able to estimate the scope of the project correctly, build up an effective DL-based solution for the given problem, and evaluate the success of the project appropriately.

At this time, we would like to thank you for reading this book. We are excited to see this book closing the gap between theory and application in the field of AI. In the same way that we gained many insights into this domain as we composed this book, we hope that your journey with us was an exceptional learning experience.

Index

Symbols

`Packt.com`

Subscribe to our online digital library for full access to over 7,000 books and videos, as well as industry leading tools to help you plan your personal development and advance your career. For more information, please visit our website.

Why subscribe?

- Spend less time learning and more time coding with practical eBooks and Videos from over 4,000 industry professionals

- Improve your learning with Skill Plans built especially for you

- Get a free eBook or video every month

- Fully searchable for easy access to vital information

- Copy and paste, print, and bookmark content

Did you know that Packt offers eBook versions of every book published, with PDF and ePub files available? You can upgrade to the eBook version at `packt.com` and as a print book customer, you are entitled to a discount on the eBook copy. Get in touch with us at `customercare@packtpub.com` for more details.

At `www.packt.com`, you can also read a collection of free technical articles, sign up for a range of free newsletters, and receive exclusive discounts and offers on Packt books and eBooks.

Other Books You May Enjoy

If you enjoyed this book, you may be interested in these other books by Packt:

Applied Machine Learning Explainability Techniques

Aditya Bhattacharya

ISBN: 978-1-80324-615-4

- Explore various explanation methods and their evaluation criteria
- Learn model explanation methods for structured and unstructured data
- Apply data-centric XAI for practical problem-solving
- Hands-on exposure to LIME, SHAP, TCAV, DALEX, ALIBI, DiCE, and others
- Discover industrial best practices for explainable ML systems
- Use user-centric XAI to bring AI closer to non-technical end users
- Address open challenges in XAI using the recommended guidelines

Deep Learning with fastai Cookbook

Mark Ryan

ISBN: 978-1-80020-810-0

- Prepare real-world raw datasets to train fastai deep learning models
- Train fastai deep learning models using text and tabular data
- Create recommender systems with fastai
- Find out how to assess whether fastai is a good fit for a given problem
- Deploy fastai deep learning models in web applications
- Train fastai deep learning models for image classification

Packt is searching for authors like you

If you're interested in becoming an author for Packt, please visit `authors.packtpub.com` and apply today. We have worked with thousands of developers and tech professionals, just like you, to help them share their insight with the global tech community. You can make a general application, apply for a specific hot topic that we are recruiting an author for, or submit your own idea.

Share Your Thoughts

Now you've finished *Production-Ready Applied Deep Learning*, we'd love to hear your thoughts! Scan the QR code below to go straight to the Amazon review page for this book and share your feedback or leave a review on the site that you purchased it from.

https://packt.link/r/1-803-24366-X

Your review is important to us and the tech community and will help us make sure we're delivering excellent quality content.